Lecture Notes in Mathematics 1948

Editors:
J.-M. Morel, Cachan
F. Takens, Groningen
B. Teissier, Paris

Frédéric Cao · José-Luis Lisani
Jean-Michel Morel · Pablo Musé
Frédéric Sur

A Theory of Shape Identification

 Springer

Frédéric Cao
DxO Labs
3 rue Nationale
92100 Boulogne Billancourt, France
fcao@dxo.com

José-Luis Lisani
Dep. Matemàtiques i Informàtica
University Balearic Islands
ctra. Valldemossa km.7,5
07122 Palma de Mallorca, Balears
Spain
joseluis.lisani@uib.es

Jean-Michel Morel
CMLA, Ecole Normale
Supérieure de Cachan
61 av. du Président Wilson
94235 Cachan Cédex, France
morel@cmla.ens-cachan.fr

Pablo Musé
Instituto de Ingeniería Eléctrica
Facultad de Ingeniería
Julio Herrera y Reissig 565
11300 Montevideo, Uruguay
pmuse@fing.edu.uy

Frédéric Sur
Loria Bat. C - projet Magrit
Campus Scientifique - BP 239
54506 Vandoeuvre-lès-Nancy Cédex
France
sur@loria.fr

ISBN 978-3-540-68480-0   e-ISBN 978-3-540-68481-7
DOI 10.1007/978-3-540-68481-7

Lecture Notes in Mathematics ISSN print edition: 0075-8434
                    ISSN electronic edition: 1617-9692

Library of Congress Control Number: 2008927359

Mathematics Subject Classification (2000): 62C05, 62G10, 62G32, 62H11, 62H15, 62H30, 62H35, 68T10, 68T45, 68U10, 91E30, 94A08, 94A13, 94B70

Cover design: WMX Design Bender

Printed on acid-free paper

9 8 7 6 5 4 3 2 1

springer.com

Preface

Recent years have seen dramatic progress in shape recognition algorithms applied to ever-growing image databases. They have been applied to image stitching, stereo vision, image retrieval, image mosaics, solid object recognition and video and web shape retrieval. More fundamentally, the ability of humans and animals to detect and recognize shapes is one of the enigmas of perception. Digital images and computer vision methods open new ways to address this enigma.

Given a dictionary of digitized shapes and a previously unobserved digital image, the aim of shape recognition algorithms is to know whether some of the shapes in the dictionary are present in the image. This book describes a complete method that starts from a query image and an image database and yields a list of the images in the database containing the query shapes.

Technically speaking there are two main issues. The first is extracting invariant shape descriptors from digital images. Indeed, a shape can be seen from various angles and distances and in various lights. A shape can even be partially occluded by other shapes and still be identifiable. Because the extraction step is so crucial, three acknowledged shape descriptors, SIFT (Scale-Invariant Feature Transform), MSER (Maximally Stable Extremal Regions) and LLD (Level Line Descriptor) will be introduced.[1]

The second issue is deciding whether two shape descriptors are identifiable as the *same shape* or not. This decision process will derive from a unique paradigm, called the Helmholtz principle. For each decision a background model is introduced. Then one decides whether an event of interest (such as the presence of a shape in the image) has occurred if it has a very low probability of occurring by chance in the background model. Thus from the statistical viewpoint shape identification goes back to *multiple hypothesis testing*.

A shape descriptor is recognized if it is not likely to appear by chance in the background model. At a higher complexity level, a group of shape descriptors is recognized if its spatial arrangement could not occur just by chance. These two decisions

---

[1] In a recent review paper on affine invariant recognition written by a pool of experts, SIFT and MSER were actually acclaimed as the best shape descriptors [122].

rely on simple stochastic geometry and eventually compute a false alarm number for each shape descriptor. The lower this number, the more secure the identification. In that way most familiar simple shapes or images can be reliably identified. Many realistic experiments show false alarm rates ranging from 10^{-5} to less than 10^{-300}.

All in all these lecture notes prove that many shapes can indeed be identified. For these shapes one needs no *a priori* model and no training, just one sample of the shape and what statisticians call a *background model*, or *a null model*. In the case of shape recognition, the term background is to be taken to the letter. By the Helmholtz principle a shape is conspicuous if and only if it cannot be generated by the image background on which it is perceived. The background model is therefore easily learnt from the image database itself.

The above description should not be taken to suggest that the shape recognition problem is solved. The methods described only apply to solid shapes and not to deformable shapes. They only deal with individual shapes and images such as logos or paintings, and not with wide classes of objects such as all humans, all cats or all cars. This latter problem is known as *categorization* and is still widely open to research.

The authors are indebted to their collaborators for many important comments and corrections, particularly to Andrés Almansa, Yann Gousseau and Guoshen Yu. David Mumford and another anonymous referee made valuable comments which reshaped the book. All experiments were done using the public software MegaWave (authors: Jacques Froment and Lionel Moisan). The SIFT method is also public and downloadable.

The present theory was mainly developed at the Centre de Mathématiques et Leurs Applications, at ENS Cachan, at the Universitat de les Illes Balears and at IRISA, Rennes. It was partially financed for the past eight years by the Centre National d'Etudes Spatiales, the Centre National de la Recherche Scientifique, the Office of Naval research (grant N00014-97-1-0839) and the Ministère de la Recherche (project ISII-RNRT), and the Ministerio de Educación y Cultura (project MTM2005-08567). Special thanks to Bernard Rougé and Wen Masters for their great interest and support. We are indebted to Nick Chriss for numerous stylistic corrections.

Frédéric Cao
José Luis Lisani
Jean-Michel Morel
Pablo Musé
Frédéric Sur

Contents

1 Introduction 1
  1.1 A Single Principle 1
  1.2 Shape Invariants and Consequences 4
      1.2.1 Shape Distortions 4
  1.3 General Overview 9
      1.3.1 Extraction of Shape Elements 9
      1.3.2 Shape Element Encoding 11
      1.3.3 Recognition of Shape Elements 11
      1.3.4 Grouping 12
      1.3.5 Algorithm Synopsis 12

Part I Extracting Image boundaries

2 Extracting Meaningful Curves from Images 15
  2.1 The Level Lines Tree, or Topographic Map 15
  2.2 Matas *et al.* Maximally Stable Extremal Regions (MSER) 17
  2.3 Meaningful Boundaries 18
      2.3.1 Contrasted Boundaries 18
      2.3.2 Maximal Boundaries 19
  2.4 A Mathematical Justification of Meaningful Contrasted
      Boundaries 21
      2.4.1 Interpretation of the Number of False Alarms 21
  2.5 Multiscale Meaningful Boundaries 26
  2.6 Adapting Boundary Detection to Local Contrast 27
      2.6.1 Local Contrast 29
      2.6.2 Experiments on Locally Contrasted Boundaries 30
  2.7 Bibliographic Notes 32
      2.7.1 Edge Detection 32
      2.7.2 Meaningful Boundaries vs. Haralick's Detector 33
      2.7.3 Level Lines and Shapes 34
      2.7.4 Tree of Shapes, FLST, and MSER 35
      2.7.5 Extracting Shapes from Images 35

Part II Level Line Invariant Descriptors

3   Robust Shape Directions 41
   3.1   Flat Parts of Level Lines 41
          3.1.1   Flat Parts Detection Algorithm 42
          3.1.2   Reduction to a Parameterless Method 43
          3.1.3   The Algorithm 44
          3.1.4   Some Properties of the Detected Flat Parts 44
   3.2   Experiments 45
          3.2.1   Experimental Validation of the Flat Part Algorithm 45
          3.2.2   Flat Parts Correspond to Salient Features 47
   3.3   Curve Smoothing and the Reduction of the Number
          of Bitangent Lines 49
   3.4   Bibliographic Notes 54
          3.4.1   Detecting Flat Parts in Curves........................ 54
          3.4.2   Scale-Space and Curve Smoothing..................... 57

4   Invariant Level Line Encoding................................ 61
   4.1   Global Normalization and Encoding 61
          4.1.1   Global Affine Normalization 61
          4.1.2   Application to the MSER Normalization Method.......... 64
          4.1.3   Geometric Global Normalization Methods 65
   4.2   Semi-Local Normalization and Encoding...................... 66
          4.2.1   Similarity Invariant Normalization and Encoding Algorithm 67
          4.2.2   Affine Invariant Normalization and Encoding Algorithm ... 70
          4.2.3   Typical Number of LLDs in Images 71
   4.3   Bibliographic Notes 73
          4.3.1   Geometric Invariance and Shape Recognition............. 73
          4.3.2   Global Features and Global Normalization 74
          4.3.3   Local and Semi-Local Features........................ 75

Part III Recognizing Level Lines

5   *A Contrario* Decision: the LLD Method 81
   5.1   *A Contrario* Models 81
          5.1.1   Shape Model or Background Model?.................... 81
          5.1.2   Detection Terminology 83
   5.2   The Background Model 85
          5.2.1   Deriving Statistically Independent Features
                   from Level Lines 87
   5.3   Testing the Background Model 89
   5.4   Bibliographic Notes 91
          5.4.1   Shape Distances 91
          5.4.2   *A Contrario* Methods 92

6   Meaningful Matches: Experiments on LLD and MSER 93
   6.1   Semi-Local Meaningful Matches 93
      6.1.1   A Toy Example 94
      6.1.2   Perspective Distortion 98
      6.1.3   A More Difficult Problem 101
      6.1.4   Slightly Meaningful Matches between Unrelated Images . . . 104
      6.1.5   Camera Blur 105
   6.2   Recognition Relative to Context 113
   6.3   Testing *A Contrario* MSER (Global Normalization) 116
      6.3.1   Global Affine Invariant Recognition. A Toy Example 116
      6.3.2   Comparing Similarity and Affine Invariant Global
            Recognition Methods 116
      6.3.3   Global Matches of Non-Locally Encoded LLDs 118

Part IV   Grouping Shape Elements

7   Hierarchical Clustering and Validity Assessment 129
   7.1   Clustering Analysis 129
   7.2   *A Contrario* Cluster Validity 131
      7.2.1   The Background Model 131
      7.2.2   Meaningful Groups 132
   7.3   Optimal Merging Criteria 136
      7.3.1   Local Merging Criterion 136
   7.4   Computational Issues 140
      7.4.1   Choosing Test Regions 140
      7.4.2   Indivisibility and Maximality 142
   7.5   Experimental Validation: Object Grouping Based
      on Elementary Features 143
      7.5.1   Segments 144
      7.5.2   DNA Image 146
   7.6   Bibliographic Notes 148

8   Grouping Spatially Coherent Meaningful Matches 151
   8.1   Why Spatial Coherence Detection? 151
   8.2   Describing Transformations 153
      8.2.1   The Similarity Case 153
      8.2.2   The Affine Transformation Case 154
   8.3   Meaningful Transformation Clusters 155
      8.3.1   Measuring Transformation Dissimilarity 155
      8.3.2   Background Model: the Similarity Case 157
   8.4   Experiments 158
   8.5   Bibliographic Notes 161

9   Experimental Results 167
   9.1   Visualizing the Results 167
   9.2   Experiments 168
        9.2.1   Multiple Occurrences of a Logo 168
        9.2.2   Valbonne Church 173
        9.2.3   Tramway 175
   9.3   Occlusions 177
   9.4   Stroboscopic Effect 179

Part V The SIFT Method

10   The SIFT Method 185
   10.1 A Short Guide to SIFT Encoding 185
        10.1.1 Scale-Space Extrema 186
        10.1.2 Accurate Key Point Detection 187
        10.1.3 Orientation Assignment 188
        10.1.4 Local Image Descriptor 189
        10.1.5 SIFT Descriptor Matching 189
   10.2 Shape Element Stability versus SIFT Stability 190
        10.2.1 An Experimental Protocol 190
        10.2.2 Experiments 191
        10.2.3 Some Conclusions Concerning Stability 195
   10.3 SIFT Descriptors Matching versus LLD *A Contrario* Matching 196
        10.3.1 Measuring Matching Performance 198
        10.3.2 Experiments 201
   10.4 Conclusion 207
   10.5 Bibliographic Notes 207
        10.5.1 Interest Points of an Image 207
        10.5.2 Local Descriptors 207
        10.5.3 Matching and Grouping 208

11   Securing SIFT with *A Contrario* Techniques 209
   11.1 *A Contrario* Clustering of SIFT Matches 209
   11.2 Using a Background Model for SIFT 210
   11.3 Meaningful SIFT Matching 214
        11.3.1 Normalization 214
        11.3.2 Matching 215
        11.3.3 Choosing Sample Points 218
   11.4 The Detection Algorithm 219
        11.4.1 Experiments: Securing SIFT Detections 220
   11.5 Bibliographic Notes 224

A Keynotes 225
   A.1 Cluster Analysis Reader's Digest 225
      A.1.1 Partitional Clustering Methods 225
      A.1.2 Iterative Methods for Partitional Clustering 227
      A.1.3 Hierarchical Clustering Methods 228
      A.1.4 Cluster Validity Analysis and Stopping Rules 231
   A.2 Three classical methods for object detection based on spatial
      coherence 235
      A.2.1 The Generalized Hough Transform 235
      A.2.2 Geometric Hashing 236
      A.2.3 A RANSAC-based Approach 237
   A.3 On the Negative Association of Multinomial Distributions 239

B Algorithms 243
   B.1 LLD Method Summary 243
   B.2 Improved MSER Method Summary 244
   B.3 Improved SIFT Method Summary 245

References 247

Index 255

Chapter 1
Introduction

1.1 A Single Principle

Digital images became an object of scientific interest in the seventies of the last century. The emerging science dealing with digital images is called *Computer Vision*. Computer Vision aims to give wherever possible a mathematical definition of visual perception. It can be therefore viewed in the realm of perception theory. Images are, however, a much more affordable object than percepts. Indeed, digital images are sampled real or vectorial functions defined on a part of the plane, usually a rectangle. They are accessible to all kinds of numerical, geometric, and statistical measurements. In addition, the results of artificial perception algorithms can be confronted to human perception. This confrontation is both advantageous and dangerous. Experimental results may easily be misinterpreted during visual inspection. The results look disappointing when matched with our perception. Obvious objects are often very hard to find in digital images by an algorithm.

In a recent book by Desolneux et al. [54], a general mathematical principle, the so called Helmholtz principle, was extensively explored as a way to define all visual percepts (gestalts) as large deviations from randomness. According to the main thesis of these authors one can compute detection thresholds deciding whether a given geometric structure is present or not in each digital image. Several applications of this principle have been developed by these authors and others for the detection of alignments [50], boundaries [51, 35], clusters [53, 33], smooth curves and good continuations [30, 31], vanishing points [2] and robust point matches through epipolar constraint [128].

These works make extensive use of a computed function, the so called *number of false alarms* (NFA) of a perception. The NFA of a perception is the expected number of times this perception could have arisen just by chance in the background. An observed configuration in an image can be numerically defined as a perception if and only if its NFA is smaller than 1. Experimental evidence has confirmed that the NFA of many human percepts of geometric figures is actually much smaller than 1,

F. Cao et al., *A Theory of Shape Identification*. Lecture Notes in Mathematics 1948.
© Springer-Verlag Berlin Heidelberg 2008

typically less than 10^{-n} where n ranges from 10 to 100 and more. See [52] and the textbook [54].

Thus theory and experiments give a mathematical and experimental basis to the existence of sure percepts. Their existence had been stated for a long time by phenomenology, in particular the Gestalt theory [95], but without quantitative evidence.

The idea that perceptions are objects unlikely to form just by chance in a background goes back to Helmholtz [83]. This principle could not be tested until images became digital and therefore accessible to computational experiments. Before the above-mentioned works, there had been several attempts to define percepts as exceptional events. Stewart [168] proposed to detect planes in a cloud of points by what he called the MINPRAN method. He computed the probability that MINPRAN will "hallucinate a fit where there is none". This probability was computed under the *a contrario* assumption that the points were randomly distributed. Lowe [112] proposed a detection framework based on the computation of accidental occurrence.

> *In other words, we can shift our attention from finding properties with high prior expectations to those that are sufficiently constrained to be detectable among a realistic distribution of accidentals.[...] Even when we do not know the ultimate interpretation for some grouping and therefore its particular* a priori *expectation, we can judge it to be significant based on the non-accidentalness criteria.*

Later Barlow [17] interpreted the perceptual goals of the neocortex as a search for *suspicious coincidences*. In the same spirit, Grimson and Huttenlocher [75, 76] proposed to compute shape recognition thresholds from a null model viewed as *the conspiracy of random*.

There is a common sense objection to applying the Helmholtz principle to shape recognition. Watching the sky, one often sees castles, cats and dogs in the clouds. Humans have a high capacity for hallucinating familiar generic shapes such as faces in rich visual environments. Thus the Helmholtz principle is not suited for all sorts of shapes.

The situation is, however, quite different regarding more specific iconic shapes such as letters, logos and in general solid shapes. One sees faces in the sky, but certainly not this or that particular face. It is to be expected that any complex enough solid shape will be recognizable in the Helmholtz sense: no random arrangement would be able to reproduce it accurately.

The mathematics to prove this are quite simple. Let $\mathcal{S}$ and $\mathcal{S}'$ be two shapes observed in two different images and which happen to be similar. Denote their (small) Hausdorff distance after registration by $\delta = d(\mathcal{S}, \mathcal{S}')$. Assume we know enough of the background model to compute the probability $\Pr(\mathcal{S}, \delta) = \Pr(d(\mathcal{S}, \Sigma) \leqslant \delta)$ that some shape in the background, Σ be as similar to $\mathcal{S}$ as $\mathcal{S}'$ is. If this probability is very small one can deduce that $\mathcal{S}'$ does not look like $\mathcal{S}$ just by chance. Then $\mathcal{S}$ and $\mathcal{S}'$ will be *identified as the same shape*.

Digital images contain thousands of significant *shape elements* that constitute their shape contents. (Several kinds of *shape elements, or descriptors* will be considered in this book.) Controlling the number of wrong matches involves the computation of the probability of a casual match with the background. This probability should be very small. But also the number of tests, which can be huge, must be under control.

Definition 1. Let $\mathcal{I}$ and $\mathcal{I}'$ be two images and N, N' the number of shape elements in each. Let $\mathcal{S}$ and $\mathcal{S}'$ be two shape elements extracted from I and I' respectively, lying at distance δ. We call number of false alarms of the match between $\mathcal{S}$ and $\mathcal{S}'$ the number

$$\mathrm{NFA}(\mathcal{S}, \mathcal{S}') = N \cdot N' \cdot \mathrm{Pr}(d(\mathcal{S}, \Sigma) \leqslant \delta).$$

If $\mathrm{NFA}(\mathcal{S}, \mathcal{S}')$ is much smaller than 1, one deduces that $\mathcal{S}'$ could not look like $\mathcal{S}$ just by chance and concludes that $\mathcal{S}$ and $\mathcal{S}'$ have *the same shape*.

There is an important phenomenological consequence: Shapes can be defined without learning, that is without empirical knowledge. By definition a shape is any part of an image which has been identified (in the sense of low NFA) at least once in another image.

From an empirical point of view, there are two kinds of shapes. First, any solid physical object can be photographed under many views and illuminations. If, by using the above definition, two snapshots of the same physical object happen to contain recognizable shape elements, one may say that the object itself is identifiable. These shape elements will constitute the object signature.

Second, humans build all kinds of standardized objects. Likewise, two different standard objects can be identified if they stem from the same industrial process. This also applies to the very numerous iconic planar shapes generated by human visual communication, in particular characters and logos. In the experimental parts of this book, we shall study the identifiability of several such iconic shapes: the lower the NFAs they generate at a given Hausdorff distance, the more recognizable they are.

As a consequence of the present study, one can define solid shapes as equivalence classes of recognized pairs without reference to empirical knowledge or *ground truth*. Thus, one should demonstrate the existence of, say, the Coca-Cola logo just by the fact that a certain group of shape elements appears in several images with very low NFA for all pairwise comparisons. Experiments will compare several snapshots of the same painting or poster, various images extracted from the same movie, or various logos of the same firm. The aim in all cases is to single out and group in clusters all shape elements common to both images. Conversely, the same method gives a negative answer when two images have no shape in common. In that case the NFA is above 1, which means that the shape is likely to occur casually in the background.

From the mathematical and numerical point of view, the main challenge in the whole study is the accurate computation of numbers of false alarms (NFA). This requires the computation of very small probabilities. Small probabilities cannot be directly measured from a shape database as frequencies. Thus, a probabilistic model of the set of all possible shapes should be built. Such a realistic experimental *background model* should be made of a large and representative set of all kinds of digital images. Unfortunately, there is no available probabilistic model for a large set of images. It is as hopeless as building a global model of the world. Even if such a model were available, one would still face the challenge of computing accurately the probability of very rare events in this world model.

Fortunately enough it is possible to overcome or rather to circumvent these two obstacles. The only information needed is the probability for a background shape to be very close to a given query shape. By a geometric independence argument, this probability will be made into a product of much larger probabilities. These probabilities instead become observable as frequencies in a small image database.

1.2 Shape Invariants and Consequences

1.2.1 Shape Distortions

In order to find the shape invariance classes, it suffices to give a rough typology of the transformations that affect images but not our recognition of the shapes they contain. Following Lisani *et al.* [109], the main classes of perturbations which do not affect recognition are:

1. **Changes in the color and luminance scales (contrast changes).** According to Gestaltists Attneave [13] and Wertheimer [179], shape perception is independent of the gray level scale or of measured colors.

   The concentration of information in contours is illustrated by the remarkable similar appearance of objects alike in contour and different otherwise. The "same" triangle, for example, may be either white on black or green on white. Even more impressive is the familiar fact that an artist's sketch, in which lines are substituted for sharp color gradients, may constitute a readily identifiable representation of a person or thing. Attneave, 1954.

   I stand at the window and see a house, trees, sky.
   Theoretically I might say there were 327 brightnesses and nuances of color. Do I have "327"? No. I have sky, house, and trees. It is impossible to achieve "327" as such. And yet even though such droll calculation were possible and implied, say, for the house 120, the trees 90, the sky 117 – I should at least have this arrangement and division of the total, and not, say, 127 and 100 and 100; or 150 and 177. Wertheimer, 1923.

   Refer to Fig. 1.1 designed by E. H. Adelson for a striking illustration of illumination invariance.

2. **Occlusions and background modification.** Shape recognition can also be performed in spite of occlusion and varying background, as shown in Fig. 1.2. The phenomenology of occlusion was thoroughly studied by Kanizsa [95] who argues that occlusion is always present in every day vision: most objects are partially hidden by others. Human perception must therefore be able to recognize partial shapes. Conversely, if a shape occludes a background, its recognition is invariant to changes in the background. This independence of shape

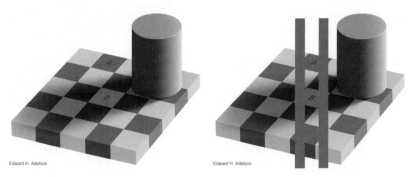

Fig. 1.1 Contrast change invariance. In the left hand image, the A and B squares have exactly the same gray level. This incredible fact is easily checked in the right hand image where A and B are linked by two rectangles with the same gray level. This experiment by E.H. Adelson illustrates the unreliability of brightness perception and the invariance of shape recognition with respect to illumination changes. (Courtesy E.H. Adelson, http://web.mit.edu/persci/people/adelson/checkershadow_illusion.html)

recognition from its background is known in perception psychology as the *figure-background problem* (Rubin [153]). The figure-background problem is part of the occlusion problem. A shape is superimposed on a background, which can be made of various objects. How can the shape be singled out from that clutter? This poses a dilemma. Extract the shape and then recognize it or extract it *because* it has been recognized?

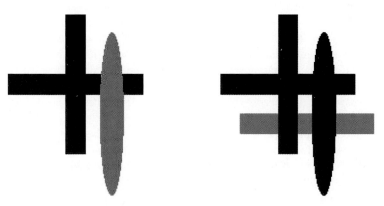

Fig. 1.2 Left: According to Kanizsa and his school, shapes can be recognized even when they undergo several occlusions. Our perception is trained to recognize shapes which are only seeable in part. Here the occluded cross can be easily recovered. Right: the *figure-background* problem. Our perception is adapted to recover a figure on the foreground, independently from the background

3. **Classical noise and blur**, inherent to any perception task and to any image generated according to Shannon's theory.

4. **Geometrical distortions or deformations.** Perspective is deeply incorporated in human perception. Humans can recognize objects and shapes under perspective distortion as long as perspective is not too strong. Recognition is also invariant to elastic deformations, always within some limits.

The previous four invariant properties fix requirements a good image representation should comply with. It will be necessary to formulate a mathematical model for each of them and to derive a well adapted image representation.

1. a. **The local contrast invariance requirement.** A digital image is usually defined as a function $u(x)$, where $u(x)$ represents the gray level or luminance at x. The first task is to extract from the image topological information independent from the varying and unknown contrast change function of the optical or biological apparatus. One can model such a contrast change function as any continuous increasing function g from $\mathbb{R}^+$ to $\mathbb{R}^+$. The real datum corresponding to the observed u could be as well any image $g(u)$. This simple argument can lead to select the level sets of the image [161] or its set of level lines as a complete contrast invariant image description [37]. If u is of class say C^1, then the level lines are the connected components of $u^{-1}(\lambda)$, which are C^1 curves for almost every $\lambda \in \mathbb{R}$. This is the choice adopted by the LLD (level line descriptor) and MSER (maximally stable extremal regions) methods. Another way to handle the contrast invariance requirement is to encode only the direction of the gradient of u, $\frac{Du}{|Du|}$ and not the gradient Du itself. Indeed, the direction of gradient is normal to the level line and is not altered by any increasing contrast change. This is the way adopted by the SIFT method to cope with contrast changes.

    b. **The concentration of information requirement.** Somewhat in contradiction to this contrast invariance principle, the Gestaltist Attneave [13] asserted that *"Information is concentrated along contours (i.e., regions where color changes abruptly)"*. Indeed not all the level lines are needed to have a complete shape description. Most of them are due to noise or to tiny illumination changes. Thus, it makes sense to select only the most contrasted level lines. That is to say, those along which the gradient of u is large enough. Such a selection is not invariant to all contrast changes, since it explicitly uses the gradient value. However, it is still invariant to affine global contrast changes. Figure 1.3 shows an example of such level lines selection. The selection of the most contrasted level lines will be the subject of Chap. 2. It will be applied to the LLD and the MSER methods. The SIFT methods actually weights its gradient orientation histograms by the gradient magnitude (see Chap. 10).

2. **The occlusion and figure-background requirements.** Even the best adapted choice of level lines is not totally suited to describing image parts. Indeed, when a shape A partially occludes a shape B, the level lines of the resulting image

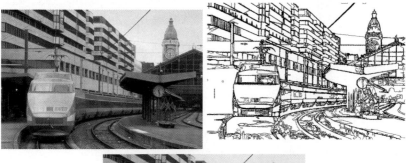

Fig. 1.3 Top left: original image, 83,759 level lines. Top right: meaningful boundaries (883 level lines). Bottom: reconstruction from the meaningful boundaries. Only 883 boundaries remain. The structure of the image is preserved and perceptual loss is very weak. The LLD and MSER methods use these boundaries for building up normalized shape descriptors. See Chap. 4

are a concatenation of pieces of the level lines belonging to A and to B. This is shown with a very simple example in Fig. 1.4. Even if a shape is not occluded, but simply occludes its own background, there may be no level line surrounding the whole shape, as shown in Fig. 1.5. These remarks show that whole level lines are too big and too sensitive to occlusion. In order to overcome this obstacle the general idea is to build shape recognition on shape elements as local as possible. The SIFT method takes small image patches. The LLD method splits level lines into small pieces.

3. **The smoothing requirement.** It is an easy experiment to check that shapes are easily recognized in images subject to noise. This means that shape information is not affected by noise. Noise introduces details which are too fine (in relation to the essential shape information) to be perceptually relevant in terms of recognition. Quoting Attneave (ibid., 1954):

> *It appears, then, that when some portion of the visual field contains a quantity of information grossly in excess of the observer's perceptual capacity, he treats those components of information which do not have redundant representation somewhat as a statistician treats "error variance", averaging out particulars and abstracting certain statistical homogeneities.*

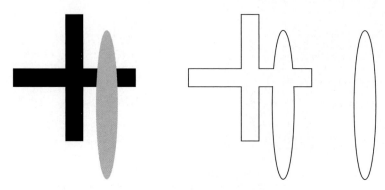

Fig. 1.4 Left: oval occluding a cross. Right: the level lines of the resulting image. While the boundary of the oval can be recovered as a full level line, the boundary of the cross concatenates with the oval boundary. Thus recognition cannot be based on complete level lines, but it can still be based on pieces of level lines, as made by the LLD method

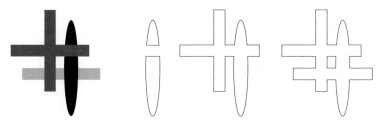

Fig. 1.5 Left: cross on a background with an oval occluding a rectangle. The cross is wholly in view. All the same, its shape does not appear as a single level line because of the background. As in Fig. 1.4, one sees that the level lines must be broken into pieces to get clues of each single shape

Hence, a correct image representation, which does not get lost in textural details and noise requires a previous smoothing. This fact is illustrated by Fig. 1.6. The object on the right was obtained by smoothing the one on the left. Both objects differ in their small details. Nevertheless, most people would recognize a black disk on both sides. Thus most shape recognition methods perform some sort of smoothing. The SIFT method applies the heat equation (which is not affine invariant) and the LLD method uses the affine scale space equation.

4. **Geometric invariance requirements.** Image representations (a set of meaningful level lines, for instance) have to be invariant to weak projective transformations. Allowing invariance to any projective transformation does not make sense, since one cannot recognize shapes under strong perspectives. Besides, it can be shown that all planar curves within a large class can be mapped arbitrarily close to a circle by projective transformations. This result was reported by Åström in [12], where it is also shown that given a finite set of m Jordan curves $C_1, \ldots, C_m$, one can find a Jordan curve C and m projective transformations $p_1, \ldots, p_m$, such that $p_i(C)$ is arbitrarily close to C_i, for all $i \in \{1, \ldots, m\}$.

Fig. 1.6 One can immediately see that both objects are disks with approximately the same size. The second one is obtained from the first by the affine curvature equation [5]. The main ideas behind such a curvature equation were anticipated by Attneave, who proposed to smooth silhouettes by blurring and then enhancing the resulting image to get a smooth silhouette: "somewhat as if the photograph of the object were blurred and then printed on high-contrast paper"

Hence, in general, schemes based on projective normalization of Jordan curves are not possible. Another argument against general projective invariance is that, despite some interesting attempts [63], there is no practical way to define a projective invariant local smoothing. From this viewpoint, affine invariant smoothing is the best compromise [5].

1.3 General Overview

The considerations on identification of Sect. 1.1 should not overshadow the other aspects of shape modeling discussed in Sect. 1.2. It is a general agreement that four main tasks must be performed properly on digital images to realize shape identification: *extraction, encoding, recognition*, and *grouping*. We shall review them in turn and see how they can be performed in a way matching all invariance requirements.

1.3.1 Extraction of Shape Elements

The first task is to define the shape elements to be compared. Indeed, images are not compared globally, but detail-to-detail and up to several geometric and photometric perturbations which can alter them drastically. In the huge amount of raw information contained in a digital image, one therefore has to define the invariant features which will become shape characteristics. This part of the programme will be accomplished by carefully translating several psychophysical invariance laws in mathematical and numerical terms. Following the Gestalt invariance laws proposed by Wertheimer, Attneave and Kanizsa [179, 13, 95], Sect. 1.2 has shown that the *shape elements* involved in shape recognition must be

1. invariant to contrast changes;
2. independent of the viewpoint, and therefore covariant by a subgroup of the projective group;
3. insensitive to the noise inherent to any image acquisition device;
4. robust to partial occlusions, and therefore local enough;
5. robust to sub-sampling, which is nothing but a zoom out.

1.3.1.1 LLD and MSER Descriptors

The invariance and robustness requirements led at least two groups of researchers to the conclusion that shape elements must be obtained from contrasted enough pieces of image level lines. Indeed, level lines are contrast invariant, covariant to any image deformation. The well contrasted level lines are also moderately sensitive to noise. The MSER method [118] uses the so called *Maximally Stable Extremal Regions* whose boundaries are nothing but well contrasted level lines. The LLD method [109, 139, 140] uses well contrasted pieces of level lines. This makes the method more insensitive to occlusions and other local shape perturbations. In the LLD method level lines are adequately smoothed by an affine invariant smoothing process. Chap. 2 deals with the extraction of LLD (or MSER) and displays many experiments. The numerical challenge is to extract as few level lines as possible from an image with no loss in shape contents.

To summarize, LLDs satisfy the requirements 1-4, but not the zoom invariance 5. We shall see in Sect. 10.2.2.1 how to make them zoom-invariant too. Level line affine smoothing is only addressed briefly in Sect. 3.3. This section explains why the robustness to noise and the invariance properties single out the affine curve scale space as the best multiscale representation of a planar curve.

1.3.1.2 SIFT Descriptors

We shall compare LLD and MSER to SIFT, another popular shape descriptor. SIFT stands for Scale Invariant Feature Transformation. SIFT descriptors are local image patches computed by the following steps:

- Simulate by Gaussian convolutions followed by sub-samplings various image zoom-outs;
- Extract scale and similarity covariant feature points at all zoom factors;
- Build a contrast invariant code of a local patch around each feature point. The size of the local patch is proportional to its zoom-out scale.

More details will be given in Chap. 10. It is easy to check that SIFT descriptors satisfy the requirements 1 and 3-5 of Sect. 1.3.1. As for the second requirement, SIFT descriptors are clearly similarity covariant but not affine covariant.

These considerations explain why we actually compared two main methods and did not bring up just one. There is no method making good all five requirements. It is not even clear that there will ever be one, because some compromise between the incompatible five requirements must be met. For instance a full affine invariance increases the size of the shape elements and decreases the locality. Also, affine transformations do not commute with zoom-ins, because affine transformations do not commute with convolutions.

In summary the LLD method matches all requirements except one and the SIFT method matches all requirements except one. Fortunately enough we will see first that both methods eventually lead to comparable results. In fact, the main identification tools developed in this book apply to all mentioned recognition methods, as demonstrated in Chap. 10 and 11.

1.3.2 Shape Element Encoding

After the extraction of shape elements has been performed the second task in view is the invariant geometric encoding of the shape elements, be them pieces of level lines or local patches. For the level lines (MSER and LLD), this is a tricky geometric computational issue. It is treated in Part II. This part describes procedures to locally encode the level lines to obtain LLDs, namely Level Lines Descriptors. Such local encoding must be robust with respect to partial occlusions.

The main step in local encoding is the choice of intrinsic local frames and scales associated with each shape element. Chap. 3 describes a way to compute stable directions for LLDs. These directions are then used in Chap. 4 to extract similarity or affine covariant shape elements. The local invariant frames permit to build for each LLD a normalized affine shape representative which can be directly compared to other shape codes. Such methods eliminating the effect of a similarity or an affine transformation are called normalization methods. The geometric normalization of Chap. 4 will prove much more robust than classical moment normalization methods. A normalization must also be performed for SIFT patches: choice of dominant directions in the patch and similarity invariant and contrast invariant encoding of the patch. This process is described in Chap. 10.

1.3.3 Recognition of Shape Elements

The third task and the main object of these notes is *identification*. This step is crucial, and usually the Achilles' heel of shape recognition methods. Part III is fundamental, short though it is. It aims to answer the question: Are two given shape elements meaningfully alike? The probabilistic modeling of the background model for LLDs is given in Chap. 5. It is tested in Chap. 6. For the SIFT method, a background model and the computation of NFAs are developed in Chap. 11.

1.3.4 Grouping

Using at first only local shape elements makes a recognition method robust to partial occlusions. A whole shape is defined as a set of shape elements in a particular geometric configuration to each other. The construction of these sets is the object of Part IV. Very interestingly, the problem can be formulated in terms of data point clustering. Clustering is one of the main techniques of Pattern Recognition. Chap. 7 focuses on the problems of cluster validation and of stopping rules in hierarchical clustering methods. These rules are applied to the grouping of shape elements in Chap. 8 for LLD and 11 for SIFT. Chap. 9 discusses many experiments.

1.3.5 Algorithm Synopsis

In Appendix A are gathered some short tutorials on the references used to write these lectures. Elements of comparison between the methods presented here and more classical ones are also developed. RANSAC, Hough transform, and geometric hashing will be considered. The keynotes can be read independently from the main text.

Since each chapter introduces tools which are all used in the shape recognition process, it seems useful to give a synopsis of the whole algorithm. Sect. B.1 presents such a synopsis for LLD, Sect. B.2 does the same for MSER and Sect. B.3 for SIFT.

Part I
Extracting Image boundaries

Chapter 2
Extracting Meaningful Curves from Images

Abstract The set of level lines of an image (isophotes) or topographic map is a complete and contrast invariant representation of an image. Level lines are ordered by inclusion in a tree structure. These two structure properties make level lines excellent candidates to shape representatives. However, some complexity issues have to be handled: The number of level lines in eight-bits encoded images of size 512×512 is typically 10^5. Most of them are very small lines due to noise or micro-texture. So the stable level lines must be selected, namely the ones that are likely to correspond to image contours. The starting point is the MSER method, a variant of the Monasse and Guichard Fast Level Set Transform. The MSER selects a set of level lines which are local extrema of contrast. This method will be put in the Helmholtz framework, following the *a contrario* boundary detection algorithm by Desolneux, Moisan and Morel [51], [54] and two powerful recent variants. The experiments in this chapter will show that selecting the most meaningful level lines reduces their number by a factor 100 without significant shape contents loss.

A method which selects one out of hundred level lines in the image without significant information loss is necessarily sophisticated. Sect. 2.1 briefly reviews the level line tree of a digital image. Sect. 2.2 describes a first way to extract well contrasted level lines, the MSER method. Sect. 2.3 makes an account of the Desolneux et al. maximal meaningful boundaries and Sect. 2.4 gives a mathematical justification which was actually missing in the original theory. Sect. 2.5 is devoted to a multi-scale extension which avoids missing boundaries because of high noise level and Sect. 2.6 deals with the so called "blue sky" effect which can lead to over-detections in textured parts of the image.

2.1 The Level Lines Tree, or Topographic Map

A gray level digital image u_d is a function defined in a rectangular grid that takes values in a finite set, typically integer values between 0 and 255. Such a datum must be interpolated to obtain a grid independent representation. According to Shannon's

F. Cao et al., *A Theory of Shape Identification*. Lecture Notes in Mathematics 1948.   15
© Springer-Verlag Berlin Heidelberg 2008

theory, this interpolation must be band-limited and is therefore analytical. Simpler spline interpolation methods can provide interpolates of arbitrary regularity class C^k for any $k \in \mathbb{N}$. All of these interpolates yield for $k \geqslant 1$ level lines with a simple topological structure.

The following theorem makes use of the so-called Jordan curves. Let us give a definition of this notion, which will come back in the sequel.

Definition 2. A Jordan curve is a simple closed curve, i.e. a closed curve that does not self-intersect.

Theorem 1. *Let u be a C^1 image in $\mathbb{R}^2$. Then, for almost all $\lambda \in \mathbb{R}$ and for all compact domain $K \subset \mathbb{R}^2$, the set $K \cap \partial(u^{-1}[\lambda, +\infty))$ is a finite set of C^1 Jordan curves. These curves called the level lines of u are either closed or meet the boundary of K at exactly two points.*

These facts are easy consequences of the Sard and implicit functions theorems [126]. The *topographic map* of an image, defined as the set of all of its level lines, is a complete representation of an image and satisfies two main properties:

- It is invariant with respect to contrast changes. Indeed, if g is an increasing function from $\mathbb{R}$ to $\mathbb{R}$, then u and $g(u)$ have the same level lines (up to a set of levels with measure 0).
- Level lines do not meet each other and are organized in a tree structure by inclusion.

Numerically, the continuous image is produced by a bilinear interpolation (order 1 spline) and is therefore not C^1. Yet it is easily checked that at almost all levels the level lines are Jordan curves and piecewise C^1. So the above-mentionned structure properties still hold. Among the possible interpolations, the bilinear interpolation presents two advantages: it is the most local of all continuous spline interpolations and does not create new extrema in the image. There is no need to compute the level lines at too many levels. It is in practice enough to take all levels $n + \frac{1}{2}$ where n goes from 0 to 255. This choice minimizes grid effects, as illustrated in Fig. 2.1.

Fig. 2.1 Left: level lines from the piecewise bilinear interpolated image. The quantization step for the gray levels is 10 starting from 0. Some grid effect (pixelization) can be seen. Right: level lines from the piecewise bilinear interpolated image with a gray level quantization step of 10 starting at gray level 0.5. These level lines show less pixelization effects

2.2 Matas *et al.* Maximally Stable Extremal Regions (MSER)

Prior to the use of level lines, shape analysis was performed in Mathematical Morphology by associating with any image a family of binary images obtained by thresholding the image at all levels and taking the upper level connected components of these binary images. This yields a complete representation of the image by its so called upper level sets. The (upper) level set of a gray level image $u : \mathbb{R}^2 \to \mathbb{R}$ at level λ is defined by

$$\chi_\lambda(u) = \{x \in \mathbb{R}^2, \quad u(x) \geqslant \lambda\}.$$

An image can be reconstructed from the whole family of its level sets, by the straightforward formula

$$u(x) = \sup\{\lambda \in \mathbb{R}, \quad x \in \chi_\lambda(u)\}.$$

The *level lines* are obtainable by simply taking the boundaries of upper level sets. The tree structure of the topographic map will be extensively used in the sequel to build an efficient computational representation of the level lines. The MSER method is a variant of the above principles for shape extraction, which go back to the seventies. We refer to the bibliographical notes for a detailed genealogy. MSER or *Maximally Stable Extremal Regions* are nothing but a selection of the most robust connected components of upper and lower level sets. This variant was introduced by J. Matas et al. in [118] in the following terms:

> In most images there are regions that can be detected with high repeatability since they possess some distinguishing, invariant and stable property. We argue that such regions (...) may serve as the elements to be put into correspondence either in stereo matching or object recognition.

Extremal regions is the name given by the authors to the connected components of upper or lower level sets. Maximally stable extremal regions, or MSER, are defined as maximally contrasted regions in the following way. Let $Q_1, ..., Q_{i-1}, Q_i, ...$ be a sequence of nested extremal regions, i.e. $Q_i \subset Q_{i+1}$ where Q_i is defined by a threshold at level i or, in other terms, Q_i is an upper (resp. lower) level set at level i. An extremal region Q_{i_0} in the list is said to be maximally stable if the area variation $q(i) =: |Q_{i+1} \setminus Q_{i-1}|/|Q_i|$ has a local minimum at i_0, where $|Q|$ denotes the area of a region $|Q|$. Clearly the above measure is a measure of contrast along the boundary ∂Q_i of Q_i. Indeed, assuming that u is C^1 and that the gray level increment between i and $i + 1$ is infinitesimal, the area $|Q_{i+1} \setminus Q_{i-1}|$ varies least when $\int_{\partial Q_i} |\nabla u|$ is maximal. It is a straightforward consequence of their definition that the MSER regions possess the robustness and invariance properties listed in Chap. 1.2. More precisely:

- Invariance to every affine transformation of image intensities;
- Covariance to all image transformations which preserve area up to a multiplicative constant. This includes affine maps;

- Stability, since only extremal regions whose support is virtually unchanged over a range of thresholds are selected;
- Multi-scale detection. Since no smoothing is involved, both very fine and very large structure are detected.

The third property makes MSER fit to undergo an *affine normalization* permitting the quick retrieval of similar deformed shapes across several images. Their affine normalization is described in Chap. 4.

The rest of this chapter describes more precise contrast based level line detection methods. Indeed, as is obvious from their definition, MSERs obey a local maximum contrast requirement which can lead to the detection of unreliable regions. Because of complexity issues, it is crucial to be able to retain all and only the relevant boundaries in each image for further shape matching.

2.3 Meaningful Boundaries

This section addresses the problem of selecting the most contrasted level lines in an image, as in the psychophysical theory of Attneave [13]. This selection involves at least two measurements, namely the length of the level line and its contrast. Intuitively, short and very contrasted level lines should be retained and less contrasted and longer ones as well. Short and non-contrasted level lines should be ruled out as unstable and therefore irrelevant. The correct weighting of length and contrast in the decision is the main object of the Desolneux et al. [51] theory. These authors proposed an *a contrario* method according to which a level line is a meaningful boundary if it could not appear in noise. We shall explain this theory in the next section and improve it in the rest of the chapter.

2.3.1 Contrasted Boundaries

Let $u : \mathbb{R}^2 \to \mathbb{R}$ be a differentiable gray level image[1]. Contrast at each point is computed as the norm of the image gradient. In order to detect level lines of u with *unexpectedly* high contrast an *a contrario* hypothesis must be proposed, under which the observed contrast on the level line will be unlikely. In the *a contrario model* contrast values are random independent identically distributed variables at all level line samples. The contrast law in this *a contrario* model is learned from the image, and approximated by its empirical histogram. So in the *a contrario* model the gradient norm follows the law of the positive random variable X defined by

---

[1] If u is a bilinearly interpolated image, then it is Lipschitz continuous and piecewise C^1. Thus its gradient is a L^∞ function, defined everywhere except on the mesh linking the center of pixels, which is a negligible set.

$$\forall \mu > 0, \quad \Pr(X > \mu) = \frac{\#\{x \in \Gamma, \, |Du(x)| > \mu\}}{\#\{x \in \Gamma, \, |Du(x)| > 0\}}, \tag{2.1}$$

where the symbol $\#$ denotes the cardinality of a set, Γ the finite sampling grid and the gradient Du is computed by a finite difference approximation. In the following, the inverse repartition function is denoted by

$$H_c(\mu) = \Pr(|Du| > \mu).$$

Definition 3 ([51], [54]). Let N_{ll} be the number of level lines of u. A level line C with length l is an ε-meaningful boundary if

$$\mathrm{NFA}(C) \equiv N_{ll} H_c(\min_{x \in C} |Du(x)|)^{l/2} < \varepsilon, \tag{2.2}$$

This quantity is called the *number of false alarms* (NFA) of C.

In (2.2) the number of false alarms or NFA is the product of the number of level lines by the probability that a random curve Γ containing $\frac{l}{2}$ independent samples has its contrast larger than $\min_{x \in C} |Du(x)|$ everywhere. If the NFA is too small the *a contrario* assumption must be rejected and we get an *a contrario* detection. Notice that meaningful boundaries are not invariant with respect to all contrast changes. Indeed, let $g : \mathbb{R} \to \mathbb{R}$ be a C^1 function and $v = g(u)$. Let C be a level line of u with level λ and $\mu = \min_{x \in C} |Du(x)|$. Then C is a level line of v with level $g(\lambda)$. Since $|Dv| = |g'(u)||Du|$ the lines change contrast depending on g and can lose or gain meaningfulness.

An interesting exception occurs with affine contrast changes. If $g(s) = as + b$, then $|Dv| = |a| \cdot |Du|$. Hence, if $a \neq 0$, the inverse repartition function of the norm of $|Dv|$ is $H'(\mu) = H_c\left(\frac{\mu}{|a|}\right)$. Therefore

$$H'(\min_{x \in C} |Dv(x)|) = H'(|a| \min_{x \in C} |Du(x)|) = H_c(\min_{x \in C} |Du(x)|),$$

and the number of false alarms of C in v is the same as in u. This proves the following result.

Lemma 1. *Meaningful boundaries are invariant to affine contrast changes.*

2.3.2 Maximal Boundaries

As Desolneux *et al.* [51] saw, meaningful boundaries usually appear in parallel groups because of the blur inherent to well-sampled images. In order to eliminate the redundancy of contrasted boundaries, these authors use the inclusion tree structure described in Sect. 2.1. Using the standard terminology of trees (nodes, branches, leaves), remember that the nodes of the tree are the level lines of the image. The ordering is defined by inclusion. This means that a Jordan level line C_1 is the parent of another one C_2 if it surrounds it, and there is no other one surrounding C_2 and surrounded by C_1. The leaves of the tree are the level lines which do not surround other ones.

Definition 4 ([134]). A monotone section of a tree of level lines is a part of a branch such that each line has a unique child and where the gray level is monotone (no contrast reversal).
A maximal monotone section is a monotone section which is not strictly included in another one.

Definition 5 ([51]). A level line is maximal meaningful if its NFA is minimal in a maximal monotone section of the level line tree.

Figure 2.2 shows how negligible the information loss is when representing an image by its maximal meaningful boundaries. They represent roughly one hundredth of all level lines.

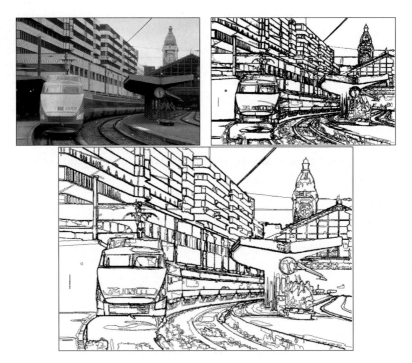

Fig. 2.2 Maximal meaningful boundaries. Top left: original image, 83,759 level lines. Top right: all meaningful boundaries, 11,505 detections. Bottom: maximal meaningful boundaries; only 883 boundaries remain, almost no detail is lost

Since maximal meaningful boundaries inherit the tree structure of the tree of level lines, they can be used to reconstruct an image (see Fig. 2.3).

Fig. 2.3 Left: original image, 99,829 level lines. Right: reconstruction from the 394 maximal meaningful boundaries. The gray level is constant and equal to the average image gray level in each connected component of the complementary of these level lines. Notice how the main shape features are preserved, while textures are removed. This simplification and reconstruction algorithm is obtained from a pruning of the tree of level lines. Salembier and Serra [154] call such operators *connected operators*

2.4 A Mathematical Justification of Meaningful Contrasted Boundaries

In this section Def. 3 is discussed. It will be shown that this definition does not prevent meaningful level lines from containing parts with low contrast. A simple cleaning rule will be derived to remove these parts.

2.4.1 Interpretation of the Number of False Alarms

The following classical lemma will be used several times in this book.

Lemma 2. *Let X be a real random variable. Let $F(x) = \Pr(X \leqslant x)$ be the repartition function of X. Then, for all $t \in (0,1)$*

$$\Pr(F(X) < t) \leqslant t.$$

In the same way, let $H(x) = \Pr(X \geqslant x)$. Then for all $t \in [0,1]$,

$$\Pr(H(X) < t) \leqslant t.$$

Proof. Let us define the pseudo-inverse

$$F^{-1}(t) = \inf\{s,\ F(s) \geqslant t\}. \tag{2.3}$$

Because of the convention in its definition, F is right-continuous. Hence

$$F \circ F^{-1}(t) \geqslant t.$$

Moreover, for all $x \in \mathbb{R}$,

$$F(x) < t \Leftrightarrow x < F^{-1}(t). \tag{2.4}$$

Indeed, let us first assume that $F(x) < t$. If $x \geqslant F^{-1}(t)$, then, since F is nondecreasing, we have $F(x) \geqslant F \circ F^{-1}(t) \geqslant t$, which is a contradiction. Conversely, let us assume $x < F^{-1}(t)$. Then, $F(x) \geqslant t$ would contradict the definition of $F^{-1}(t)$. This proves the equivalence. Hence,

$$\begin{aligned}
\Pr(F(X) < t) &= \Pr(X < F^{-1}(t)) \quad \text{by (2.4)} \\
&= \Pr(\exists y, \, y < F^{-1}(t), \, X \leqslant y) \\
&= \sup_{y < F^{-1}(t)} F(y) \\
&\leqslant t \quad \text{again by (2.4).}
\end{aligned}$$

The third equality is a basic convergence theorem of measure theory. Note that the last inequality is not strict, because of the passage to the limit. The second part of the lemma is proved in the same way by introducing $H^{-1}(t) = \sup\{s, \, H(s) \geqslant t\}$. The proof is left to the reader. □

Let us remark that if F is continuous and increasing, F^{-1} is really the inverse of F and the lemma then yields an equality, and means that $F(X)$ is a uniform variable in $(0, 1)$.

The *A Contrario* Model

Let us assume that X is a real random variable described by the inverse repartition function $H(\mu) = \Pr(X \geqslant \mu)$. Assume that u is a random image such that the values of $|Du|$ at each point in the sample grid are independent, and follow the same law as X. Let now E be a set of random curves (C_i) in u such that $\#E$ (the cardinality of E) is independent of each C_i. For each i, let $\mu_i = \min_{x \in C_i} |Du(x)|$. Let us also assume that L_i independent points can be chosen on C_i. The curves C_i can be thought of as random walks with independent increments but since a finite number of samples are selected on each curve, the law of the C_i does not really matter. Let us finally assume that L_i is independent from the pixels crossed by C_i. Such a random model is called *a contrario* or *background* model.

A curve C_i is said to be ε-meaningful if

$$\mathrm{NFA}(C_i) = \#E \cdot H(\mu_i)^{L_i} < \varepsilon.$$

Remark 1. In digital images, the independence of the values of $|Du|$ is sound only if the points are far enough from each other. In practice, the minimal distance will be taken equal to 2, since a 2×2 finite difference scheme is used to compute the image gradient.

The following proposition justifies Def. 3.

Proposition 1. *The expected number of ε-meaningful curves in a random set E of random curves is smaller than ε.*

Proof. Let us denote by X_i the binary random variable equal to 1 if C_i is meaningful and to 0 else. Let also $N = \#E$. Let us denote by $\mathbb{E}(X)$ the expectation of a random variable X in the *a contrario* model. Then

$$\mathbb{E}\left(\sum_{i=1}^{N} X_i\right) = \mathbb{E}\left(\mathbb{E}\left(\sum_{i=1}^{N} X_i \mid N\right)\right).$$

It is assumed that N is independent from the curves. Thus, conditionally to $N = n$, the law of $\sum_{i=1}^{N} X_i$ is the law of $\sum_{i=1}^{n} Y_i$, where Y_i is a binary variable equal to 1 if $nH(\mu_i)^{L_i} < \varepsilon$ and 0 else. By linearity of the expectation,

$$\mathbb{E}\left(\sum_{i=1}^{N} X_i \mid N = n\right) = \mathbb{E}\left(\sum_{i=1}^{n} Y_i\right) = \sum_{i=1}^{n} \mathbb{E}(Y_i).$$

Since Y_i is a Bernoulli variable, $\mathbb{E}(Y_i) = \Pr(Y_i = 1) = \Pr(nH(\mu_i)^{L_i} < \varepsilon) = \sum_{l=0}^{\infty} \Pr(nH(\mu_i)^{L_i} < \varepsilon \mid L_i = l)\Pr(L_i = l)$. Again, it is assumed that L_i is independent of the gradient distribution in the image. Thus conditionally to $L_i = l$, the law of $nH(\mu_i)^{L_i}$ is the law of $nH(\mu_i)^{l}$. Let us finally denote by $(\alpha_1, \cdots, \alpha_l)$ the l (independent) random values of $|Du|$ along C_i. Then,

$$\Pr\left(nH(\mu_i)^{l} < \varepsilon\right) = \Pr\left(H(\min_{1 \leqslant k \leqslant l} \alpha_k) < \left(\frac{\varepsilon}{n}\right)^{1/l}\right)$$

$$= \Pr\left(\max_{1 \leqslant k \leqslant l} H(\alpha_k) < \left(\frac{\varepsilon}{n}\right)^{1/l}\right) \quad \text{since } H \text{ is nonincreasing}$$

$$= \prod_{k=1}^{l} \Pr\left(H(\alpha_k) < \left(\frac{\varepsilon}{n}\right)^{1/l}\right) \quad \text{by independence}$$

$$\leqslant \frac{\varepsilon}{n} \quad \text{from Lemma 2.}$$

This term does not depend upon l. Thus

$$\sum_{l=0}^{\infty} \Pr(nH(\mu_i)^{L_i} < \varepsilon \mid L_i = l)\Pr(L_i = l) \leqslant \frac{\varepsilon}{n}\sum_{l=0}^{\infty}\Pr(L_i = l) = \frac{\varepsilon}{n}.$$

Hence,

$$\mathbb{E}\left(\sum_{i=1}^{N} X_i | N = n\right) \leqslant \varepsilon.$$

This finally implies $\mathbb{E}\left(\sum_{i=1}^{N} X_i\right) \leqslant \varepsilon$, which means exactly that the expected number of ε-meaningful curves is less than ε. $\square$

In this proposition, it is not assumed *a priori* that the C_i are level lines of u. Indeed, in this case it cannot be asserted that the length (number of independent points) of the curve is independent from the values of the gradient along the curve.

2.4.1.1 Cleaning up Meaningful Boundaries

In the rest of the chapter we will deal with (necessary) improvements of the original Desolneux et al. method. Proposition 1 asserts that if a curve is a meaningful boundary, then it cannot be *entirely* generated in white noise (up to ε false detections on average). On the other hand, can it be guaranteed that no part of a meaningful boundary is contained in noise? Or, for a given meaningful boundary, is it possible to give an upper bound of the size of the part of the boundary that is likely to be contained in noise (i.c. a non-edge region)? To answer this question, let us introduce the *a posteriori* length distribution

$$p_\mu(l) \equiv \Pr(L \geqslant l | \min_{x \in C} |Du(x)| \geqslant \mu), \tag{2.5}$$

where the image model is a white noise, C is a level line in this image, and L its length. Remark that for small values of μ, the position of level lines is inaccurate. Indeed, in a region where $|Du| < 1$ uniformly, for any two points x and y such that $|u(x) - u(y)| = 1$, then $|x - y| > 1$: Level lines cannot be too close. Hence, given a reasonable value of μ (for instance 1), any small piece of curve with a point of gradient less than μ will be removed from the meaningful level line. The next question is "how small"? Following the same philosophy as for meaningful level lines, let us now consider an image u with N_{ll} (quantized) level lines. Let us also denote by N_l the number of all possible sampled subcurves of these level lines. (If the curves are closed with L_i independent points, then $N_l = \sum_{i=1}^{N_{ll}} L_i^2$.) Let us assume that a piece of curve with length l does not contain any point with gradient less than μ. If l satisfies $N_l \cdot p_\mu(l) < \varepsilon$, then we can consider that this piece of curve cannot be due to chance. On the contrary, if l does not satisfy this inequality, the curve is more likely to have been generated by noise. The decision is to remove the latter curves. More precisely, the algorithm is:

1. Detect meaningful boundaries.
2. For a fixed $\mu > 0$, let $\mathcal{L}(\mu) = \inf\{l, N_l \cdot p_\mu(l) < \varepsilon\}$.
3. For any meaningful boundary, remove every subcurve of length $\mathcal{L}(\mu)$ containing a point where $|Du| \leqslant \mu$.

There is a last problem though: how to compute $p_\mu(l)$? This requires the distribution law of the length of level lines in a noise image. It can be estimated empirically, for short enough level lines. For instance, in a 500×500 image, there are many level lines with a length less than about 1000, and the distribution is considered as accurately approximated by the empirical distribution. (See Fig. 2.4.) Then, using Bayes' rule,

$$p_\mu(l) = \frac{\sum_{k=l}^{\infty} \Pr(\min_{x \in C} |Du(x)| \geqslant \mu | L = k) \Pr(L = k)}{\sum_{k=1}^{\infty} \Pr(\min_{x \in C} |Du(x)| \geqslant \mu | L = k) \Pr(L = k)}.$$

By independence of the gradient values along the level lines,

$$p_\mu(l) = \frac{\sum_{k=l}^{\infty} H_c(\mu)^k \Pr(L = k)}{\sum_{k=1}^{\infty} H_c(\mu)^k \Pr(L = k)}. \tag{2.6}$$

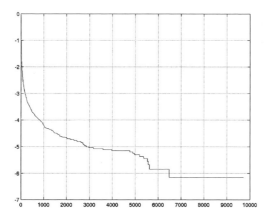

Fig. 2.4 Log10 of the inverse repartition function of length of level lines in a white noise image. The average length is about 3.5, meaning that most level lines enclose a single pixel

A new parameter, μ, has been introduced. When μ gets larger, $\mathcal{L}(\mu)$ decreases, so that the clean up procedure removes more numerous but smaller pieces of curves.

The choice of μ could depend on the application. Detected edges may be used for different purposes, for instance shape recognition or image matching. If $|Du|$ is less than 1, then the position of level lines may be locally inaccurate. Eliminating pieces of curves with a gradient smaller than $\mu = 1$ for all images is therefore not restrictive in shape recognition applications. Figure 2.5 shows an example of the clean up procedure.

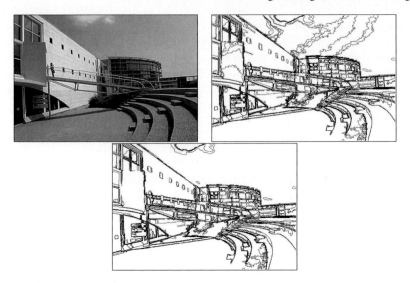

Fig. 2.5 Meaningful boundary clean up. Top left: original image. Top right: meaningful boundaries with local histograms. See Sect. 2.6. Boundaries are found in the sky. They are detected, because the gradient in the sky is regular due to the smoothly changing illumination. The gradient value is about 0.2 in the sky, but the curves are so long that they are detected. This does not contradict our detection principle: such curves are indeed exceptional in noise, since it is very unlikely that the gradient does not attain an even smaller value on such a long curve. What is actually contradicted is our assumption that these exceptional curves do correspond to edges, no matter how small the contrast is. This assumption indeed implies that one is able to distinguish arbitrary gray level changes. This is perceptually not true. Bottom: result after the clean up procedure with a gradient threshold equal to 1

2.5 Multiscale Meaningful Boundaries

The contrast measure is an approximation of the gradient by finite differences. More precisely, Desolneux *et al.* use the scheme

$$\frac{\partial u}{\partial x} \simeq u_x(i,j) = \frac{1}{2}(u(i+1,j) + u(i+1,j+1) - u(i,j) - u(i,j+1)), \quad (2.7)$$

$$\frac{\partial u}{\partial y} \simeq u_y(i,j) = \frac{1}{2}(u(i,j+1) + u(i+1,j+1) - u(i,j) - u(i+1,j)). \quad (2.8)$$

Using a 2×2 scheme is coherent with the application of Helmholtz principle: points at distance 2 have independent values of contrast in white noise. On the other hand, this value is noise sensitive. Smoothing the image before computing the gradient would partly remove noise but would also introduce local dependencies between pixels. This would make the *a contrario* model false in smoothed white noise, and false detections could be expected. Nevertheless, the *a contrario* model still applies if the smoothed image is down-sampled at a lower rate, in conformity with Shannon's sampling theory.

The Multiscale Algorithm

Consider a set of N_s dyadic scales $\{1, 2, ..., 2^{N_s-1}\}$. For any level line C, let us denote by C^s the curve $\frac{C}{2^s}$, obtained by scaling C by a factor 2^{-s}. Let also H^s denote the empirical contrast distribution of u^s, where u^s is obtained by downsampling u by a factor 2^s in agreement with Shannon's theory. (That is to say, downsampling follows an adequate smoothing, for instance convolution with a prolate function [146].)

1. Compute the quantized level lines of u.
2. For each level line C with l independent points in u, compute μ^s, the minimal value of $|Du^s|$ over all pixels crossed by C^s. Let

$$\text{NFA}(C) = N_s \cdot N_{ll} \min_{s \in \{0, \cdots, N_s-1\}} (H^s(\mu^s))^{l/2^s}. \tag{2.9}$$

A curve C is an ε-meaningful multiscale boundary if $\text{NFA}(C) < \varepsilon$.

Thus, a curve is meaningful if and only if there exists a scale such that it is $\frac{\varepsilon}{N_s}$ meaningful in the sense of the previous section. Roughly speaking, the N_s factor is the price to pay to have the right to test several different scales. Clearly it only makes sense to consider a small number of dyadic scales (say 3 or 4, since the side of usual digital image does not much exceed 2^{10} pixels). Since the detection depends on $\log \varepsilon$ [50, 51], considering $\frac{\varepsilon}{N_s}$-meaningful boundaries at each scale can eliminate only a few lines.

The expected number of detections in a white noise image is still under control.

Proposition 2. *With the same hypotheses as in Prop. 1, the expected number of ε-meaningful multiscale boundaries in a white noise image is less than ε.*

Indeed, downsampled images are still white noise images. Together with the linearity of expectation and the proof of Prop. 1, this yields the result.

Figures 2.6 and 2.7 show the result of the multiscale algorithm on images with quantization or additive white Gaussian noise. In Fig. 2.6 the shapes are not very sharp because of motion blur and transparency. Level lines following contours are very long since they surround several objects. Moreover, the background is nearly uniform. Thus the minimal contrast value along long level lines is all the more sensitive to the gradient computation. The effect is also dramatic in the noisy image of Fig. 2.7 (Gaussian noise with standard deviation 30). Note also how the boundaries of the main objects still coincide with level lines, in spite of the very strong noise.

2.6 Adapting Boundary Detection to Local Contrast

In the *a contrario* model, the values of the gradient are random variables whose distribution is empirically estimated by using the histogram of the gradient in the whole image. The use of this global distribution yields the so-called blue sky effect.

Fig. 2.6 Influence of quantization noise on meaningful boundaries. Left: the original image is coarsely quantized and has a very low contrast. This leads to bad gradient estimation and a lot of missing detections (middle). Right: multiscale detection is less sensitive to quantization noise and gives correct detections

Fig. 2.7 Multiscale meaningful boundaries and noise. Left: image of Fig. 2.9 with an additive white Gaussian noise of standard deviation 30. Middle: meaningful boundaries. Since noise dominates the gradient distribution, only six small level lines are detected. Right: multi-scale detection using four dyadic scales. Textures are not detected, meaning that noisy textures are in this case not different enough from noise to be detected. On the other hand, main structures remain. This allows one to empirically check the stability of the topographic map in spite of a significant amount of noise

Consider an image containing two parts: a contrasted or textured one (e.g. ground) and a smooth one (e.g. sky), see Fig. 2.9. There is an empirical overdetection in the ground, and an underdetection in the sky. Indeed, the sky only contributes small values in the histogram. Thus the algorithm tends to detect any level line which is more contrasted than the sky. So nearly anything is detected in the ground. Conversely, the contrasted ground may make the detection more difficult for regions with small contrast, such as a cloudy sky. This is not coherent with human vision, which locally adapts perception to contrast.

This section addresses the local adaptiveness to contrast by modifying the meaningful boundary model. It describes a local detection algorithm which extends the global one.

2.6.1 Local Contrast

Let us assume that a closed boundary has been detected. It divides the image into two connected components: the interior and the exterior of the curve. Let us compute the empirical contrast distribution in each component. Meaningful boundaries are then detected, independently in each connected component. This procedure can now be recursively applied. Since the tree of level lines of a quantized image has a finite depth, it is clear that the detection procedure stops after a finite number of steps.

Two problems make things slightly more delicate. First, the order that is used to describe the image boundaries may have an influence. The obvious solution is to treat first the most meaningful boundary at each step.

A second problem is purely computational and involves open boundaries, that is the ones whose endpoints belong to the image border. They still cut the image into two connected components, that should be processed in the same way since there is no clear notion of interior and exterior. However, in order to make the tree structure unique, exactly one of these components is considered to be the interior one. Open boundaries are then closed following the shortest path along the border of the image. This choice is only an algorithmic one, and is arbitrary from a perceptual point of view [134]. To circumvent this lack of symmetry between both connected components, the detection is first applied to open boundaries until no new open boundaries are detected. The procedure is then applied to closed boundaries.

2.6.1.1 Local Algorithm

Let us call R_0 the root boundary, that is the (non-meaningful) boundary containing all the image. If C is a boundary, its interior is denoted by $\text{Int } C$. For an illustration of the next algorithm, we refer to Fig. 2.8.

1. Set $R \leftarrow R_0$. (Local root).
2. Set $\mathcal{M}$, the set of already stored in R meaningful boundaries. Initially, $\mathcal{M}$ is empty.
3. Let $R' \leftarrow R \setminus \cup_{C \in \mathcal{M}} \text{Int } C$.
4. Compute the histogram of $|Du|$ in R'.
5. Use this histogram and detect the maximal meaningful boundaries included in R'. Let us call total maximal boundaries, the meaningful boundaries C satisfying

$$\begin{cases} \mathrm{Int}\,(C') \subsetneq \mathrm{Int}\,(C) \Rightarrow \mathrm{NFA}(C) < \mathrm{NFA}(C') \\ \mathrm{Int}\,(C) \subset \mathrm{Int}\,(C') \Rightarrow \mathrm{NFA}(C) \leqslant \mathrm{NFA}(C'). \end{cases} \qquad (2.10)$$

The set of total maximal boundaries is denoted by $\mathcal{N}$. In other words, the boundaries in $\mathcal{N}$ have an optimal NFA, since they are more meaningful than boundaries which contain them or in which they are contained. This assumption is stronger than the maximality defined in Sect. 2.3.2 since the NFA comparison is not restricted to monotone sections. The subtree with root equal to R that remains by keeping only the boundaries in $\mathcal{N}$ has only two levels: the local root R, and $\mathcal{N}$. Since the interior of open boundaries is arbitrary, the detection of open and closed boundaries are not mixed. In practice, this means that if an open meaningful boundary C is detected, the definition of total maximal boundary (2.10) is only applied to open boundaries containing C or contained in C.

6. If $\mathcal{N} \neq \emptyset$, then new boundaries have been detected in the complementary of the already detected ones. Then,

    a. Set $\mathcal{M} = \mathcal{M} \cup \mathcal{N}$. By construction, all the closed boundaries in $\mathcal{M}$ have disjoint interiors.

    b. return to step 3.

7. If $\mathcal{N} = \emptyset$, there are no new boundaries in the local root and in the complementary of the currently detected boundaries. The search is then resumed at lower levels of the tree as follows. For any boundary $C \in \mathcal{M}$,

    a. Store C.

    b. Set $R \leftarrow C$, and $\mathcal{M} \leftarrow \emptyset$.

    c. Return to step 3.

Remark 2. Each boundary may be tested more than once. Thus, the number of false alarms has to be multiplied by the maximal number of boundary visits, which is bounded from above by the depth of the level lines tree. In fact, each detected boundary often lies in the middle of the local root, and this divides the tree depth by 2. Thus the maximal number of boundary visits is of the order of the logarithm of the initial tree depth. In practice, it never exceeds 100.

2.6.2 Experiments on Locally Contrasted Boundaries

Figure 2.9 shows the difference between the detection with a global contrast histogram and the updated local histogram. To give an idea of the magnitude of the number of false alarms, the boundary separating sky and foreground has NFA 10^{-357}. This means that such contrasted lines are expected to occur less than once in 10^{357} level lines taken from white noise images. The smaller boundaries around the opening on the top of the tower have 10^{-10} NFA.

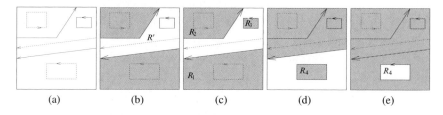

(a)   (b)   (c)   (d)   (e)

Fig. 2.8 Illustrates the local meaningful boundary algorithm. (a) the initial boundaries. They are oriented so that the tangent and the inner normal form a direct frame. The NFA of each boundary is computed. There are three total maximal boundaries (in solid line); two are open, one is closed. While some open curves are detected, the closed ones are skipped. (b) Compute the contrast histogram in the complementary set of the interior of the open detected boundaries and resume the search in this part of the image, which is the region R'. The closed boundary is total maximal meaningful again. However, no new open boundaries are detected. Thus, this closed boundary is kept. (c) The search is resumed (with recomputed histogram) in the exterior (white part) of the detected boundaries, until new ones can no longer be found, which is the case on the figure. When this is over, compute the local contrast histogram in each region R_1, R_2, R_3 and look for boundaries inside them. (d) A (closed) total maximal boundary R_4 has been detected in R_1. Compute the local histogram in $R_1 \setminus R_4$ and detect boundaries. (e) Finally, scan for boundaries in R_4 with new local contrast histogram. Since nothing is detected, the output is the boundaries of R_1, R_2, R_3 and R_4

Fig. 2.9 Influence of local contrast. From left to right: original image, maximal meaningful boundaries, local maximal meaningful boundaries. There are 280,000 boundaries in the initial image (for a gray level quantization step of 1), 652 in the second one and 193 in the last one. Texture is removed since local contrast (for instance) on the church tower is much more demanding than the global histogram. As the texture is uniform, no level line is a large deviation to the empirical local contrast, yielding no detection. This is very good for shape analysis where it is often desirable to distinguish texture from real shapes

The effect of local contrast in boundaries detection is twofold.

1. Textures are eliminated.
2. Local contrast makes curves in low contrasted areas more detectable.

This was expected since, in textured regions (as on the tower), the local contrast values are larger than in the rest of the image. Thus, this increases the NFA of boundaries; most of them simply disappear in textured regions. This is a masking phenomenon in the Gestalt terminology [95].

On the other hand, some lines are detected due to the illumination gradient (see Fig. 2.5). They can be due to the vicinity of the light source or to the variation of the orientation of the surface of a three dimensional object with respect to the light source. Such lines do not correspond to silhouettes of physical objects. Nevertheless, it is reasonable to detect them as remarkable structures.

What is the impact of the preceding study regarding shape recognition? It is well known that texture is much damaged by compression. Thus, the precise geometry of level lines in texture may depend very much on the image source (quality, compression rate, etc). Moreover, they are very complex, and will yield many encoded pieces of curves when the procedure of Chap. 4 is applied. The shape content of a texture is therefore both huge in quantity and unreliable. The computational cost to handle it may therefore be too high for some applications. Thus it may be useful to automatically remove contrasted regions corresponding to texture.

The argument above is reversed for stereo image registration or motion estimation. In this case, it is *a priori* known that the images under comparison are nearly the same image. The goal is to register them extensively. In this application, textures generate many level lines which can be tracked and should not be eliminated.

2.7 Bibliographic Notes

The presentation of this chapter follows [35], which improved the boundary detection method proposed in [51]. The next paragraphs give some genealogy for edge detection, level lines, level sets, and the topographic map.

2.7.1 Edge Detection

It is a well known fact that shape information in images is concentrated along regions where color or gray level changes abruptly [13, 116]. Since Marr and Hildreth's seminal work on edge detection [117], the effort on extracting shape information from images has been mainly concentrated on local methods. Among these methods, which are commonly referred as edge detectors, Canny [27] and Canny-Deriche [47] filters are certainly the most widely used.

Classical edge detectors have two problems. The first is that they depend on (at least) two parameters, the threshold on the contrast and the degree of smoothing. Both are hard to estimate and are usually fixed manually. The second problem with these methods is that they detect points rather than structures. The edge points have to be connected by chaining algorithms which involve further parameters.

2.7.2 Meaningful Boundaries vs. Haralick's Detector

Following Haralick [79], edges are the maxima of the gradient norm in the direction of the gradient, such that the gradient is larger than a given threshold. Thus, for a gray level image u, they are the zero-crossings of $D^2u(Du, Du)$. Since this quantity is numerically sensitive to noise, a multiscale strategy *à la* Marr is applied. In practice, u is first convolved with a Gaussian with standard deviation σ. Let us denote by g_σ this Gaussian and set $u_\sigma = g_\sigma * u$. Edge pixels are defined such that $|Du_\sigma| > \mu$ and $D^2u_\sigma(Du_\sigma, Du_\sigma)$ has a different sign for neighboring pixels. There have been some attempts to automatically determine the scale parameter σ [105], but edge detection widely remains multiscale as predicted by Marr [116]. In practice, it is quite difficult to track edges back to small scales. The multiscale meaningful boundaries detection of Sect. 2.5 allows for the consideration of various scales while keeping detection thresholds completely automatic. Moreover, the number of scales has a logarithmic influence.

A second problem is that Haralick's detector provides us with a set of points or curves containing only a few pixels. The way they should be connected is far from obvious. It may lead to a very high computational complexity and depends on several sensitive parameters. Level lines are Jordan curves, and do not have this problem.

Last but not least, Haralick's operator is inefficient for corners and junctions. Indeed, at those points, the gradient direction is very badly estimated and edges may be severely cut. Additional algorithms are necessary to reconnect pieces of edges as opposed to level lines nicely bifurcating at T-junctions and giving the different boundaries. (See Sect. 2.1.) Figure 2.10 shows the meaningful boundaries and Canny's filter (which is an optimized version of Haralick's method) near two junctions. Edges detected in that way often are very short and require additional and somewhat unreliable linking procedures. The behavior of the level lines around the T-junctions is quite clear. When extracting shape elements by local encoding, all the different configurations near the junctions will be considered. Clearly, meaningful level lines provide a set of curves which is more reliable and directly usable at the expense of a more heavy computational cost (a few seconds for a typical 512×512 images, half the time being dedicated to the computation of the level lines tree and half to the selection of meaningful boundaries).

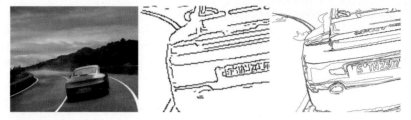

Fig. 2.10 Junction and level lines. Left: original image. Middle: Haralick's detector implemented with Canny's filter. Note how contours are broken near the junctions. Right: detailed view of meaningful boundaries on the region. Compare the accuracy of level lines on the plate with the detected edges at the same place

2.7.3 Level Lines and Shapes

Following [109], Chap. 1 asserted that the set of level lines of a digital image was a natural representation of its shape content. Indeed, it provides a geometric information invariant to contrast changes. Moreover, no chaining procedure is needed since level lines are already curves. This chapter presented briefly the bilinear level line tree proposed by Lisani *et al.* [110]. For a detailed account, see the book [54]. Whereas edge detectors usually fail near T-junctions (and additional treatments are necessary), there are several level lines at a junction (see Fig. 2.11 and [36]).

Fig. 2.11 Level lines and T-junctions. Depending on the gray level configuration between objects and background, level lines may follow or not (as on the figure) the objects boundaries. In any case, junctions appear where two level lines separate. Here, there are two kinds of level lines: those surrounding the occluded circle and those following the boundary of the union of the circle and the square. These level lines are included in each other and do not meet but are usually very close and not distinguishable along contrasted contours

2.7.4 Tree of Shapes, FLST, and MSER

Prior to the use of level lines, shape analysis was performed in Mathematical Morphology by associating with any image a family of binary images obtained by thresholding at all levels. This yields a complete representation of the image by its upper level sets [119, 161]. The tree structure of the topographic map has been extensively used to build an efficient computational representation of the level lines. See the Monasse *et al.* [135, 15, 110] algorithms. An efficient region growing algorithm, the *Fast Level Set Transform* allows one to compute the tree of level lines for digital images (constant in each pixel) or bilinearly interpolated images [110]. The idea of considering the level lines of the bilinear interpolated image was also independently proposed in the so-called Digital Morse Theory [43].

MSER stands for "maximally stable extremal regions", which are a subset of the "extremal regions" of the image. What the authors of [118] define as extremal regions are the connected components of the level sets of the image (which we call "shapes") earlier proposed by Monasse for contrast invariant image registration in [133, 134, 136]. Monasse and Guichard's shape extraction algorithm [135] is very similar to the MSER extraction. The extracted shapes are organized in a tree structure, the above-mentioned FLST. Since the set of "shapes" of an image is very big (tens of thousands of shapes can be typically found in an image of size 512×512), some selection strategy needs to be defined in order to pick the "most important" shapes. Monasse and Guichard proposed to pick the shapes with highest contrast in the shape tree, which is almost the same definition as the one given in [118]. The recent paper [56] proposed an efficient MSER algorithm for real time object tracking in video and in [138] and [166, 167] fast tree computations alternatives and variants to the FLST have also been proposed. In [51] an *a contrario* technique is used to select shapes in the level lines tree having contrasted enough boundaries. Variations of this technique are [110] and [35]. The presentation in this chapter followed the last and more sophisticated [35], which subsumes all preceding ones.

2.7.5 Extracting Shapes from Images

The extraction of shape elements is seldom addressed in the context of shape recognition. Most works on shape recognition assume that shapes are already extracted [68, 129, 151]. In Mokhtarian's approach [129, 130, 131], shapes are extracted by simply thresholding dark objects over a bright background. Their boundaries are level lines. Rothwell proposed a whole recognition system of flat objects on uniform background [151]. The shapes he treats are simply Jordan curves bounding objects. Rothwell's method builds the object boundaries by extracting edges using Canny's edge detector [28]. Canny's filter performs well in Rothwell's framework where objects are well-contrasted over a uniform background. In general, this filter is not very efficient (see 2.7.2) but in this particular easier case, one simply gets back level lines again!

Part II
Level Line Invariant Descriptors

Invariance and robustness requirements for shape recognition were discussed in Sect. 1.2. The *local contrast invariance* led us to consider image level lines. The *concentration of information requirement* required the selection of a set of *meaningful level lines* that are roughly the level lines which are long and contrasted enough (a precise definition has been given in Chap. 2). It followed from our discussion in Sect. 1.2 on *occlusion and figure-background* that *small pieces* of meaningful level lines should be considered. Finally combining the *smoothing requirement* and the *geometric invariance requirement* implies that level lines should be smoothed by an affine invariant process. This leads us to the following choice for shape elements.

Definition 6. We call *level line descriptor (LLD)* of an image any piece of well contrasted level line of the image which has been smoothed by an affine invariant smoothing process. This piece of level line will not be considered by itself, but rather by its equivalence class under all planar affine transformations.

The aim of this part is to implement the above definition of LLD. We need to explain how to segment level lines and how to find an affine invariant code for each LLD. The next chapter 3 defines robust features of level lines, the so called flat parts which will be used for affine normalization (described in Chap. 4).

Chapter 3
Robust Shape Directions

Abstract This chapter deals with shape affine normalization. This method associates with all shapes deduced from each other by an affine distortion a single normalized shape. A crucial ingredient for normalization is the computation of a small affine covariant set of robust straight lines associated with a shape. The set of all tangent lines to a shape has this covariance property, but it is too large. A very successful idea is to use bitangent lines, that is, lines tangent to a shape at two different points. If the shape has a finite number of inflexion points it also has a finite number of bitangent lines. In Sect. 3.3 a well-established curve affine invariant smoothing algorithm will be briefly described. This smoothing permits a drastic reduction of the number of bitangent lines. Yet, not all shapes can be encoded by using bitangents. Convex shapes have no bitangents and simple shapes have only a few. This explains why shape recognition algorithms compute other robust straight lines associated with the shape. Flat parts of curves are informally defined as intervals of the curve along which the direction of the tangent line does not vary too much. For instance, large enough polygons show as many reliable flat parts as sides. This chapter will present a simple parameterless definition of flat parts, based again on the Helmholtz principle.

3.1 Flat Parts of Level Lines

Flatness of a part of curve will be measured by comparing its direction at each point with the direction of the underlying chord (see Fig. 3.1).

Although flatness may look like a rather intuitive geometric concept, it is in fact quite complex. Our aim is to define a unique measurement, the flatness for very diverse phenomena: A long very oscillating curve may look flat seen at a distance. In another way, a short and very smooth curve can look locally very flat. One can therefore figure out that at least two parameters are involved in a flatness measurement. One measures the length of the flat part and the other gives the amplitude of the oscillations. Thus, the flatness definition problem can be viewed as the question

F. Cao et al., *A Theory of Shape Identification*. Lecture Notes in Mathematics 1948.   41
© Springer-Verlag Berlin Heidelberg 2008

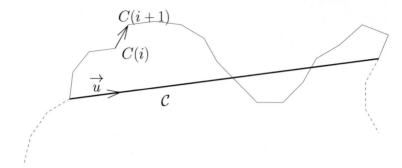

Fig. 3.1 A piece of discrete curve with the underlying chord $\mathcal{C}$ (thick segment line)

of reducing two parameters to a more abstract one, the flatness. The detection of flat parts of a curve should meet the following requirements:

- It should not detect just points around which the curve is flat, but the precise straight intervals on the curve.
- Long flat parts should be allowed to move farther from their underlying chord than short ones.
- The detection should be intrinsic to the curve, and not depend on other curves in the image.
- Detected flat parts should not overlap.
- Since detecting flat parts is generally the first step of a recognition algorithm, it deals with a huge amount of information. Therefore, computational complexity should be low.

3.1.1 Flat Parts Detection Algorithm

Consider a chord from a given curve C: its endpoints delimitate a piece of curve of length l (measured in pixels). Since one would like to measure how much the piece turns with respect to the direction $\vec{u}$ given by the chord, let us define

$$\alpha = \max_{i \in \{0 \dots n-1\}} \left\{ \left| \mathrm{angle}(\overrightarrow{C(s_i)C(s_{i+1})}, \vec{u}) \right| \right\},$$

where the discrete piece of curve is made of the n consecutive points $C(s_i)$.

Suppose that α is below some fixed threshold α^*. Following the discussion on independence in Sect. 2.5, consider that points at a geodesic distance (along the curve) larger than 2 are statistically independent. Thus, there are $l/2$ statistically independent segments of the type $(C(s_i), C(s_{i+1}))$ along a curve with length l. The probability of the event that $l/2$ statistically independent points on a piece of curve show a tangent line which makes an angle lower than α among all the pieces of curve for which $\alpha < \alpha^*$ is given by:

$$p(\alpha, l) = \left(\frac{\alpha}{\alpha^*}\right)^{l/2}.$$

Of course, the lower $p(\alpha, l)$ the flatter the piece of curve.

This straightforward computation is valid under the assumption that among all the pieces of curves such that $\alpha < \alpha^*$, α is uniformly distributed over $[0, \alpha^*]$, and that the tangents are independent at Nyquist distance 2. Flat parts are now defined as rare events with regard to this *a contrario* model.

For each piece of the curve for which $\alpha < \alpha^*$, the probability $p(\alpha, l)$ is computed. Only pieces such that $p(\alpha, l)$ is under a predetermined threshold p^* are kept (these parts are called candidates). Such pieces can of course overlap. So some of them must be selected to be the flat parts of the curves. A greedy algorithm will be used: the piece of curve with the lowest p is marked as a flat part, then all candidates that share a common part with this best flat part are eliminated. The process is iterated with the remaining candidates.

3.1.2 Reduction to a Parameterless Method

The computation of α clearly depends on the discretization. The curves which the proposed algorithm deals with are level lines of images. Their natural discretization is the pixel.

The whole algorithm involves two thresholds. The first one, α^*, is not critical. Indeed, since one is interested in detecting flat parts, it is natural to *a priori* reject all pieces of curve where α is above a large threshold. We set $\alpha^* = 1$ radian once for all, which is not a strong constraint. More specifically, a change of α^* multiplies all probabilities $p(\alpha, l)$ by a constant factor. Thus, the flatness measurement is just scaled and the ordering maintained. Moreover, changing α^* also multiplies the threshold p^* by the same constant. Thus, there are not two parameters here, but just one, namely p^*. This last parameter will be eliminated by Helmholtz principle. It can be fixed in such a way that almost no flat part occurs in the level curves of a white noise.

Experimental evidence shows that $p^* = 10^{-3}$ is the maximum value for which only a few detections (on average one) occur on level lines extracted from a white noise image containing the same amount of level lines as a standard natural image. So with this value for p^* the proposed algorithm satisfies the Helmholtz principle in that there is almost no detection of flat parts in a white noise image.

3.1.3 The Algorithm

Consider a Jordan curve on which flat parts are searched for.

Part I: Candidate identification.

For each chord of the curve with length 10, 20, 30, ..., 180, 200, and then an exponential progression[1]:

1. Compute the maximum angle α between the chord and the piece of curve delimited by both ends of the chord. If n denotes the number of independent points $C(s_i)$ on this piece of discrete curve:

$$\alpha = \max_{i \in \{1...n-1\}} \left\{ \left| \mathrm{angle}(\overrightarrow{C(s_i)C(s_{i+1})}, \overrightarrow{u}) \right| \right\}.$$

2. If $\alpha > 1$ rad, *a priori* reject the piece; else compute $p(\alpha, l) = \left(\frac{\alpha}{\alpha^*}\right)^{l/2} = \alpha^{l/2}$, where l is the length of the considered piece of curve.
3. If $p(\alpha, l) > p^* = 10^{-3}$, reject the piece.

Part II: Greedy algorithm

1. Keep the candidate for which $\alpha^{l/2}$ is minimal, mark it as *flat part*, and discard it from the list of candidates.
2. Reject all candidates that meet this best candidate.
3. Iterate until no candidate is available anymore.

3.1.4 Some Properties of the Detected Flat Parts

The condition defining the candidates ($\alpha^{l/2} < p^*$) is not a real constraint for long curves. For example, if $p^* = 10^{-3}$ and $l = 200$, all curve parts such that $\alpha < 0.97$ are accepted as candidates. Nevertheless, long pieces of curves often show subparts with a lower probability and a greedy algorithm will therefore prefer them. In the case of circles, however, this does not occur. Let us compute the arcs of circle which will be marked as flat parts. Figure 3.2 illustrates the following computations.

Proposition 3. *A circle of radius R has flat parts if and only if $R \geq -e \log(p^*)$. In such a case, the length of the detected flat parts is $L = 2R\sin(1/e)$.*

---

[1] There is a complexity issue here. All chords are not tested, but only a subsample of them so that the algorithm does not waste too much time for long curves. The only consequence of this discretization procedure is that long straight lines (in practice, lines whose length is larger than 100 pixels) can be split into two pieces (see Fig. 3.14 for an example). This is not an important drawback since the goal is to use flat parts as robust directions.

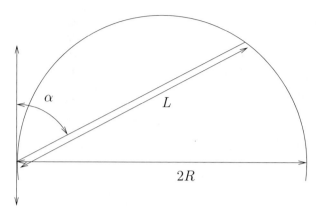

Fig. 3.2 Illustration of the flat parts computation on a circle

Proof. A circle of radius R being given, let us consider a chord of length L defining a maximum angle α with the corresponding piece of curve ($0 \leqslant \alpha \leqslant \pi/2$). The values of α and L are related by $L = 2R\sin(\alpha)$. The probability defined earlier is $p(\alpha, L) = \alpha^{R\alpha}$ (expressed as a function of L, it writes down $p(\alpha, L) = \arcsin(L/2R)^{R\arcsin(L/2R)}$). The function $\alpha \mapsto \alpha^{R\alpha}$ attains a minimum for the value $\alpha = 1/e$. Consequently, $\forall \alpha,\ \alpha^{R\alpha} \geqslant e^{-R/e}$.

Thus, if the probability threshold is set to p^*, and if $R < -e\log(p^*)$, then the circles of length R will show no flat part. On the contrary, if $R \geqslant -e\log(p^*)$, the detected flat parts (after the greedy step) in circles of radius R will always show a maximum angle $\alpha = 1/e$ (that is to say 21 degrees, corresponding to an arc of $1/9$ of the total circle), and their length will be $L = 2R\sin(1/e)$. $\square$

Notice that p^* only controls the minimum radius under which no flat part will be detected: $-e\log(p^*)$. It appears only through its logarithm and small variations of it will not influence the final result. Although for symmetry reasons no piece of circle should be favored by the algorithm, the position of the detected flat parts over a circle strongly depends on the starting point of the discrete curve describing this circle. This makes flat parts of circular curves unreliable in position. However, this will not hinder the recognition of circles, as a a circle matches well with itself, up to any rotation.

3.2 Experiments

3.2.1 Experimental Validation of the Flat Part Algorithm

Experimental results are shown in Figs. 3.4 to 3.9 (original images can be seen on Fig. 3.3). For each image, the computation time is less than 10 seconds, for a 2GHz

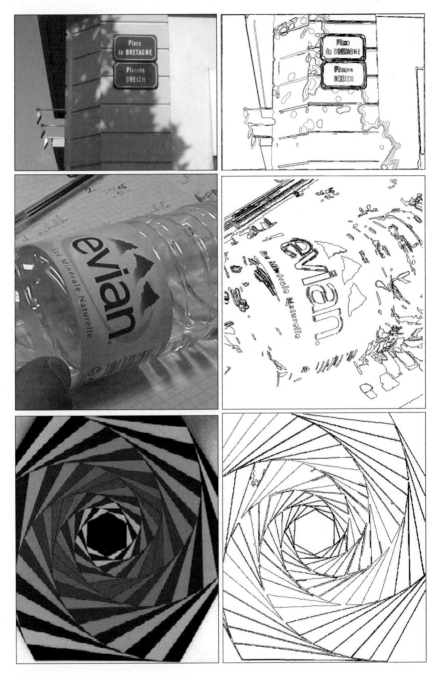

Fig. 3.3 Left column: original images. Right column: meaningful level lines detected with the method described in Chap. 2 (right). Top: *Bretagne*, 413 level lines. Middle: *Evian*, 481 level lines. Bottom: *Vasarely*, 172 level lines

standard PC. When images do not show long level lines, the computation time is less than a second.

Fig. 3.4 Flat parts detection: *Bretagne*. 1004 detections. Flat parts as small as the ones in the letters of the name of the street are detected (about 10 pixels high). Flat parts in the boundaries of the shadows can be eliminated by dropping the probability threshold, as can be seen on Fig. 3.5. Nevertheless these detections actually correspond to small flat parts

3.2.2 Flat Parts Correspond to Salient Features

Figures 3.10 and 3.11 show the result of the proposed flat parts detector over all level lines in an image. By all, we mean that all level lines at all levels with quantization step equal to 1 have been extracted. This allows for an exact reconstruction of the original image from the level lines and their corresponding gray levels [135]. Some segments are detected over level lines corresponding to quantization noise (*i.e.* not contrasted level lines over perceptually uniform areas), but these segments actually correspond to small pieces of straight lines. They are no longer detected when the probability threshold p^* is set to 10^{-10} instead of the standard value (10^{-3}). Flat parts are concentrated along edges. This experiment confirms that segment lines are actually salient image features.

Comparing Fig. 3.3 to Figs. 3.4 to 3.7 shows that almost all detected flat parts belong to maximal meaningful boundaries.

Fig. 3.5 Flat parts detection: *Bretagne*, with $p^* = 10^{-10}$, 417 detections. Letters are too small to be detected but the remaining flat parts are very accurate

Fig. 3.6 Flat parts detection: *Evian*. 448 detections

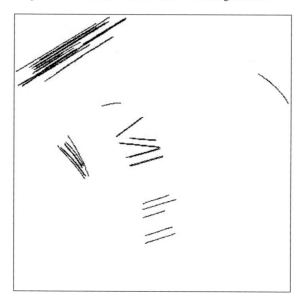

Fig. 3.7 Flat parts detection: *Evian*, with $p^* = 10^{-10}$, 64 detections

In his PhD thesis, Lisani [108] used a flat points detector to build robust semi-local normalization. Figures 3.12 to 3.15 show a comparison between the flat parts proposed in this chapter and flat points in the sense of Lisani. See captions for details.

3.3 Curve Smoothing and the Reduction of the Number of Bitangent Lines

Level lines may be subject to noise, and can have details that are too fine in relation to the essential shape information. Hence, a good shape representation requires a previous smoothing. Is this smoothing necessary? Quite, from the technological viewpoint, as otherwise there would be too many bitangent lines to level lines and therefore too many geometric codes to a level line. The general framework by which an image or a shape is smoothed at several scales in order to eliminate spurious or textural details and extract its main features is called Scale Space. The main developments of Scale Space theory in the past ten years involve invariance arguments. Indeed, a scale space will be useful for shape recognition only if it is invariant. Let us summarize a series of arguments given in [5]. A scale space computing contrast invariant information must in fact deal directly with the image level lines; in order to be local (not dependent upon occlusions), it must be in fact a partial differential equation (PDE). In order to be a smoothing, this PDE must be parabolic. The affine

Fig. 3.8 Flat parts detection: *Vasarely*, 774 detections. Each triangle side is correctly detected as a single flat part

Fig. 3.9 Flat parts detection: *Serena Williams & Puma* (original image shown in Fig. 2.6). Left: Original level lines (425 lines). Middle: $p^* = 10^{-3}$ (675 detections). Right: $p^* = 10^{-10}$ (156 detections). Flat parts on letters are correctly extracted

Fig. 3.10 Flat parts detection. (a) original image (size: 512×384); (b) 25,755 level lines (quantization step: 1 gray level). They cover the whole image. (c) 20,065 flat parts detected over these level lines (probability threshold p^* has here its standard value: 10^{-3}); (d) flat parts of length larger than 100 pixels among the previous ones; (e) 6,233 flat parts detected over these level lines, when the probability threshold p^* is set to 10^{-10}; (f) flat parts of length larger than 100 pixels among the previous ones. Flat parts appear to be concentrated along edges in thick bundles

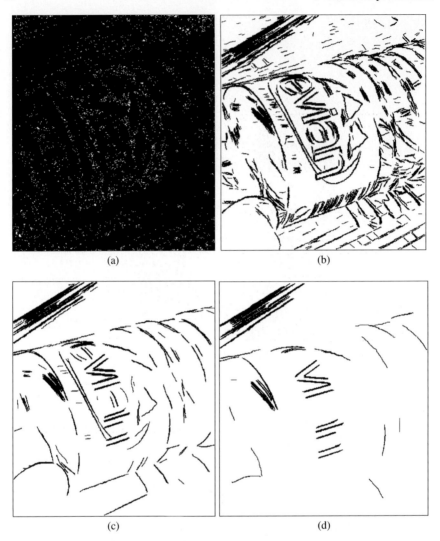

Fig. 3.11 Flat parts detection. (a) 90,078 level lines from Evian image (quantization step: 1 gray level); (b) flat parts detections over these level lines (16,533 detections); (c) flat parts detection with $p^* = 10^{-6}$ (4,659 detections); and (d) flat parts detection with $p^* = 10^{-10}$ (2,041 detections). Flat parts are concentrated along edges

Fig. 3.12 Lisani's flat points: *Serena Williams & Puma*. Only 15 flat points (in black) are detected. To be compared to the results in Fig. 3.9

invariance requirement and the invariance with respect to reverse contrast lead to a single PDE [5]. This PDE, characterizing the unique contrast, contrast reversal and special affine invariant scale space is

$$\begin{cases} \frac{\partial u}{\partial t} = |Du|(\text{curv } u)^{1/3}, \\ u(x, t) = u_0(x). \end{cases} \tag{3.1}$$

It is called *Affine Morphological Scale-Space* (AMSS). Here $u(t, 0) = u_0$ is the initial image, $u(t, x)$ is the image smoothed at scale t and $\text{curv}(u)(x) = \text{div}(\frac{Du}{|Du|})$ denotes the signed curvature of the level line passing by x. This equation is equivalent to the affine curve shortening [155] of all of the level lines of the image, given by the equation

$$\frac{\partial x}{\partial t} = |\text{Curv}(x)|^{\frac{1}{3}} \mathbf{n}, \tag{3.2}$$

where x denotes a point of a level line, $\text{Curv}(x)$ its curvature and $\mathbf{n}$ the signed normal to the curve, always pointing towards the concavity.

Moisan [127] found a fast algorithm for this curvature motion. For more details on this scheme, refer to [127, 99] and to the book [29]. The invariants mentioned mean that the evolution of a shape does not depend upon any affine distortion of the plane. This corresponds to an invariance to all orthographic projections of a planar shape.

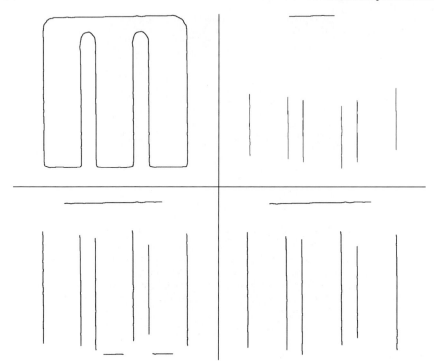

Fig. 3.13 Flat points *vs* flat parts: *Serena Williams & Puma*. From left to right and from top to bottom: considered level line, flat points (7 detections), flat parts with $p^* = 10^{-3}$ (9 detections), flat parts with $p^* = 10^{-10}$ (7 detections). One of the flat parts in the legs of the character M is not detected since these curve pieces are too small and pose a sampling problem. Since not *all* chords are tested but a subset of them, endpoints may sometimes be not conveniently distributed

Figure 3.16 shows that a slight smoothing by the affine scale space eliminates the sampling effects of a digital image and reduces drastically the number of inflexion points of a shape without altering its overall aspect. Numerically, the smoothing is slight and stops at the scale $t = 0.5$ at which a circle with radius 0.5 collapses. So the smoothing roughly eliminates details of 1 pixel size.

3.4 Bibliographic Notes

3.4.1 Detecting Flat Parts in Curves

In their seminal paper [65], Fischler and Bowles argue that any curve partitioning technique must satisfy two general principles: stability of the description, and a complete and concise explanation. Smooth sections of curves play a major role

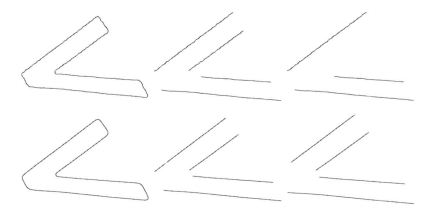

Fig. 3.14 Flat points *vs* flat parts: character V in *Evian*. Top: no smoothing. From left to right: original level line, flat parts with $p^* = 10^{-3}$ (4 detections) and with $p^* = 10^{-10}$ (3 detections). The flat points algorithm does not provide any detection. Bottom: after smoothing. From left to right: original level line, flat parts with $p^* = 10^{-3}$ (5 detections) and flat parts with $p^* = 10^{-10}$ (4 detections). With $p^* = 10^{-3}$, one of the segments is split because of the discretization procedure in the multi-scale test of chords. Again here the Lisani flat points algorithm misses the segments

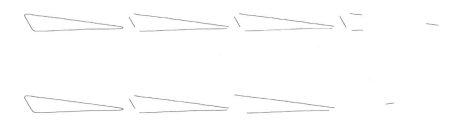

Fig. 3.15 Flat points *vs* flat parts: a triangle in *Vasarely*. Top: no smoothing. From left to right: original level line, flat parts with $p^* = 10^{-3}$ (3 detections) and flat parts with $p^* = 10^{-10}$ (3 detections), and flat points (4 detections). Bottom: after smoothing (see Sect. 3.3). From left to right: original level line, flat parts with $p^* = 10^{-3}$ (5 detections) and flat parts with $p^* = 10^{-10}$ (2 detections), and flat points (1 detection)

because they fit both principles. For instance, Guy and Medioni [78] consider segment lines as *salient* features in images. Flat part multiscale detection has been used for the more general problem of polygonal approximation of digitized curves (see [164]).

Segment or straight line detection is one of the cornerstones of computer vision. Indeed, it is often a preprocessing step of shape recognition, shape tracking [48], vanishing point detection [2], convex shape detection [92], *etc*. Most of the time,

Fig. 3.16 Some level lines of a gray level image. Quantization effects and noise are seen. After a slight smoothing these effects disappear (right)

straight lines in images are conceived as contiguous edges. Many line detection algorithms therefore require a previous local edge extraction step, such as a Canny's filtering [28]. Hough Transform [85] and algorithms derived from it [91] have been widely studied for this purpose. The goal of these methods is to identify clusters in a particular space (the parameter space of a straight line, either (ρ, θ) with ρ the distance of the line to the origin, and θ the angle between a vector normal to the line and a fixed direction, or (a, b) where a is the slope and b the ordinate of the intersection between the straight line and the ordinate axis). The Hough transform is a voting procedure: every pixel votes for the parameters of the straight line going through it. Another method consists in first chaining the local edges by taking into account connectivity (see for an example [70]), and then in identifying segments among the discrete curves [107]. The main drawbacks of these methods are the number of thresholds (edge detection needs at least a gradient threshold, and the Hough Transform needs a quantization step for the parameter space discretization and a threshold for the voting procedure) and their computational burden and instability (due to local edges chaining). A fuzzy segment concept was proposed in [45]. In this method the primary detection is still based on a set of points derived from a local edge detector.

The method presented in this chapter can be viewed as an adaptation to the level lines of Desolneux et al. [50], who proposed an *a contrario* method detecting meaningful alignments in images. A meaningful alignment is a segment where a large enough proportion of points have their gradient orthogonal to the segment. More precisely a length l segment is ε-meaningful in a $N \times N$ image if it contains at least $k(l)$ points having their direction aligned with the one of the segment, where:

- $k(l)$ is given by: $k(l) = \min\{k \in \mathbb{N}, \Pr(S_l \geqslant k) \leqslant \varepsilon/N^4\}$, and
- $\Pr(S(l) \geqslant k)$ is the probability that, in at least k points in a straight segment of length l, the gradient of the image is orthogonal to the segment, up to a predetermined precision.

Estimating the probability that k points among l have a tangent with the same direction as the chord is not relevant to detect flat parts. In such a model, consecutive alignments are indeed not favored. They are instead crucial for shape normalization.

In his PhD thesis [108], Lisani defined flat points on curves by using two arbitrary parameters. A flat point is the center of a curve segment for which the sum of the angle variations of tangents is small enough (less than 0.2 radian) over a large enough piece of curve (larger than 15 pixels). This algorithm misses many flat points, and does not really detect segments, as several experiments have shown clearly.

Figure 3.17 shows the results for some of the algorithms which were just discussed. As far as flat parts detection is concerned, Desolneux's alignments are suitable neither for detecting accurate segment directions nor for detecting segment lengths. The naive segment detector based on Hough transform which illustrates the discussion is certainly not the best that can be done using Hough techniques. Nevertheless even a more clever algorithm would face the same problem as this one. It involves numerous critical parameters (different parameters would drastically change the results). Some isolated points are detected as segments because they fall by chance on the same straight line as another more distant segment and therefore collect its votes. Both algorithms (alignments and the Hough transform-based algorithm) are not local enough: that is why segments over the characters in the test image are not detected. Canny's edge detector is well known to suffer from lack of accuracy at edge junctions (where the gradient is badly estimated). Here, this would not be a real issue, since segment lines are searched for between junctions, where edges are more accurately detected. Nevertheless those edge detectors need several critical thresholds.

3.4.2 Scale-Space and Curve Smoothing

Since the seminal work of Lamdan *et al.* [101], bitangent lines are well-known to be of high interest to build up semi-local invariant curve descriptions. The reduction of the number of bitangent lines is linked to curve smoothing, or curve scale space. The modern concept of scale space comes from Witkin [181] and is mainly related to the

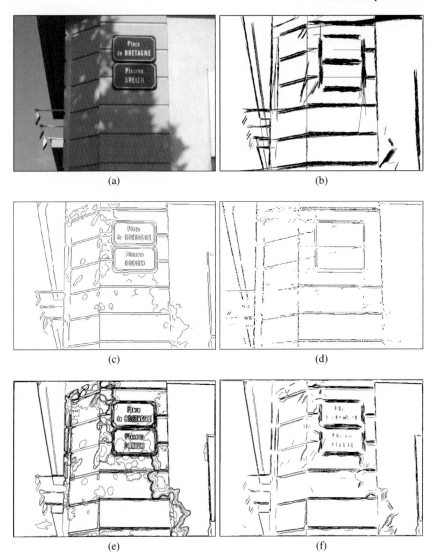

Fig. 3.17 Segment detection. (a) original image; (b) maximal meaningful alignments [50]; (c) Canny's edge detector; (d) Points that correspond to an edge and that lie at the same time on a direction detected by voting in the Hough space; (e) local maximal meaningful level lines; (f) result of the proposed algorithm. See text for discussion.

Gaussian scale space, given by the heat equation [98]. An interesting shape recognition method using the mean curvature motion was discovered by Mocktharian and Mackworth [132]. The use of curvature-based smoothing for shape analysis is by now well established. The seminal papers are [10], [132] and [62]. These authors define a multi-scale curvature which is similarity invariant, but not affine invariant. Abbasi *et al.* [1] used the mean curvature motion and an affine length parameterization of the boundary of the solid shapes in order to get an approximately affine shape encoding. Sapiro and Tannenbaum [155] and Alvarez, Guichard, Lions and Morel [5] independently discovered the affine scale space with different approaches. Alvarez *et al.* proved existence of viscosity solutions to the affine scale space. An existence and regularity theorem was later proved by Angenent, Sapiro and Tannenbaum [7] from which it can be derived that the number of inflexion points decreases under the affine scale space. This result is crucial for shape encoding. Moisan [127] found a fast and fully affine invariant scheme implementing the affine scale space. He also proved the uniform consistency, which by a Barles and Souganidis [16] result is sufficient for convergence. The numerical scheme of Moisan was later extended by Cao and Moisan [34] to more general motions by curvature. Very recently the affine erosion scheme was used by Niethammer *et al.* [142] to compute an affine invariant skeleton of plane curves.

Chapter 4
Invariant Level Line Encoding

Abstract Chapters 2 and 3 described the level lines extraction, selection and smoothing procedures, as well as the selection of a few stable, local directions on these curves. These procedures yield shape elements which cannot be directly compared or recognized since they have undergone an unknown affine transformation. The classical way to address this problem is *normalization*. We call affine invariant normalization a method to build shape representations that are invariant to any planar affine transformation $T(x) = Ax + b$, such that $\det(A) > 0$. In other words, an affine invariant normalization transforms a planar shape $\mathcal{F}$ into a normalized shape such that any deformation of $\mathcal{F}$ by a planar affine transformation will give back the same normalized shape. Notice that shapes related by axial symmetry are not considered to be equivalent in this framework and will not yield the same normalized shape. Similarity invariant normalization is simpler and will be defined in the same way. Section 4.1 first presents the most classical moment method for affine normalization. We will show that this method is not efficient. In Sect. 4.1.3, a much more accurate normalization method is proposed, involving local and robust features of a level line such as bitangent lines and flat parts. This method is applied first to global level lines and then adapted in Sect. 4.2 to pieces of level lines, thus making shape recognition robust to occlusions. These normalization techniques will be used to describe, first, the MSER moment normalization method. The more sophisticated geometric affine normalization methods will be applied throughout the book to the recognition of LLDs (level line descriptors).

4.1 Global Normalization and Encoding

4.1.1 Global Affine Normalization

Classical shape normalization methods are based on the inertia matrix normalization. We shall use Cohignac's presentation of this method [40]. This method has

F. Cao et al., *A Theory of Shape Identification*. Lecture Notes in Mathematics 1948.   61
© Springer-Verlag Berlin Heidelberg 2008

some drawbacks that are common to all moment-based normalization methods. They rely on computing high order moments and are therefore unstable and very sensitive to noise. In the next section we propose a global geometric normalization technique based on robust directions (bitangent lines and flat pieces of each level line). Thus the use of moment-based normalization is not recommended. It is, however, simple and elegant and needs to be presented before a more intricate and efficient way is proposed.

Denote by $\mathbb{1}_{\mathcal{F}}$ the indicator function of a solid shape $\mathcal{F}$. In order to achieve translation invariance of the normalized representation, it may be assumed that $\mathcal{F}$ has been previously translated so that its barycenter is at the origin of the image plane. Hence, the moment of order (p, q) (p and q natural integers) of $\mathcal{F}$ is defined by

$$\mu_{p,q}(\mathcal{F}) = \int_{\mathbb{R}^2} x^p y^q \mathbb{1}_{\mathcal{F}}(x, y) dx dy.$$

Let $S_{\mathcal{F}}$ be the following 2×2 positive-definite, symmetric matrix

$$S_{\mathcal{F}} = \frac{1}{\mu_{0,0}} \begin{pmatrix} \mu_{2,0} & \mu_{1,1} \\ \mu_{1,1} & \mu_{0,2} \end{pmatrix},$$

where $\mu_{i,j} = \mu_{i,j}(\mathcal{F})$. By the uniqueness of Cholesky factorization [71], $S_{\mathcal{F}}$ may be uniquely decomposed as $S_{\mathcal{F}} = B_{\mathcal{F}} B_{\mathcal{F}}^{\mathrm{T}}$ where $B_{\mathcal{F}}$ is a lower-triangular real matrix with positive diagonal entries.

Definition 7. The *pre-normalized shape* associated to $\mathcal{F}$ is the shape $\mathcal{F}' = B_{\mathcal{F}}^{-1}(\mathcal{F})$.

The aim is to prove that the pre-normalized solid shape is invariant to affine transformations, up to a rotation.

Lemma 3. *Let A be a non-singular 2×2 matrix. Then $S_{A\mathcal{F}} = AS_{\mathcal{F}}A^{\mathrm{T}}$.*

Proof. Let a, b, c and d be real numbers such that:

$$A = \begin{pmatrix} a & b \\ c & d \end{pmatrix}.$$

The moment of order $(2, 0)$ associated to the solid shape $A\mathcal{F}$ is

$$\mu_{2,0}(A\mathcal{F}) = \det(A) \int_{\mathbb{R}^2} (ax + by)^2 \mathbb{1}_{\mathcal{F}}(x, y) dx dy$$
$$= \det(A)(a^2 \mu_{2,0} + 2ab\mu_{1,1} + b^2 \mu_{0,2}).$$

The same computation for moments of order $(0, 2)$ and $(1, 1)$ yields

$$\mu_{0,2}(A\mathcal{F}) = \det(A)(c^2 \mu_{2,0} + 2cd\mu_{1,1} + d^2 \mu_{0,2}),$$
$$\mu_{1,1}(A\mathcal{F}) = \det(A)(ac\mu_{2,0} + bd\mu_{0,2} + (ad + bc)\mu_{1,1}).$$

Since $\mu_{0,0}(A\mathcal{F}) = \det(A)\mu_{0,0}$, one can easily check that $S_{A\mathcal{F}} = AS_{\mathcal{F}}A^{\mathrm{T}}$. □

Lemma 4. *Let X_0 be a 2×2 invertible matrix. Then, for any 2×2 matrix X: $XX^{\mathrm{T}} = X_0 X_0^{\mathrm{T}}$ if and only if there exists an orthogonal matrix Q such that $X = X_0 Q$.*

Proof. Since X_0 is invertible, $XX^{\mathrm{T}} = X_0 X_0^{\mathrm{T}}$ iff $X_0^{-1} X (X_0^{-1} X)^{\mathrm{T}} = \mathrm{Id}_2$. Letting $Q = X_0^{-1} X$ yields the result. □

Proposition 4. *The pre-normalized solid shape is invariant to any invertible, planar, linear transformation $(x, y)^{\mathrm{T}} \mapsto A(x, y)^{\mathrm{T}}$, up to an orthogonal transformation. Moreover, if $\det(A) > 0$, the invariance holds up to a rotation.*

Proof. Since A is a 2×2 non singular matrix, following Lemma 3, $S_{A\mathcal{F}} = A S_{\mathcal{F}} A^{\mathrm{T}}$. By letting $B_{\mathcal{F}}$ be the lower-triangular matrix of Cholesky's decomposition of $B_{\mathcal{F}}$, it follows that $S_{A\mathcal{F}} = A B_{\mathcal{F}} (A B_{\mathcal{F}})^{\mathrm{T}}$. Now, since $S_{A\mathcal{F}}$ is a 2×2 positive-definite, symmetric matrix, Cholesky factorization yields $S_{A\mathcal{F}} = B_{A\mathcal{F}} B_{A\mathcal{F}}^{\mathrm{T}}$, where $B_{A\mathcal{F}}$ is a 2×2 non-singular, lower-triangular real matrix. Then, by Lemma 4, $B_{A\mathcal{F}} = A B_{\mathcal{F}} Q$, where Q is a 2×2 orthogonal matrix. Hence, $B_{A\mathcal{F}}^{-1} A\mathcal{F} = (A B_{\mathcal{F}} Q)^{-1} A\mathcal{F} = Q^{-1} B_{\mathcal{F}}^{-1} A^{-1} A\mathcal{F} = Q^{-1} B_{\mathcal{F}}^{-1} \mathcal{F}$, which proves the invariance of $\mathcal{F}' = B_{\mathcal{F}}^{-1} \mathcal{F}$ to planar isomorphisms, up to an orthogonal transformation. Finally, notice that if $\det(A) > 0$, then $\det(Q) > 0$. □

A closed form for $B_{\mathcal{F}}^{-1}$ in terms of the moments of $\mathcal{F}$ can be computed by taking the inverse of $B_{\mathcal{F}}$, the lower-triangular matrix given by the Cholesky decomposition of $S_{\mathcal{F}}$,

$$B_{\mathcal{F}}^{-1} = \sqrt{\mu_{0,0}} \begin{pmatrix} 1/\sqrt{\mu_{2,0}} & 0 \\ -\mu_{1,1} \Big/ \left(\mu_{2,0} \sqrt{\mu_{0,2} - \frac{\mu_{1,1}^2}{\mu_{2,0}}} \right) & 1 \Big/ \left(\sqrt{\mu_{0,2} - \frac{\mu_{1,1}^2}{\mu_{2,0}}} \right) \end{pmatrix}.$$

The pre-normalized solid shape $\mathcal{F}' = B_{\mathcal{F}}^{-1} \mathcal{F}$ is then an affine invariant representation of $\mathcal{F}$ *modulo* a rotation. In order to obtain a full affine invariant representation, only a reference angle is needed. This can be achieved, for instance, by computing

$$\varphi = \mathrm{Arg} \left(\int_0^{2\pi} \int_0^{+\infty} \mathbb{1}_{\mathcal{F}'}(r, \theta) e^{i\theta} r \, dr \, d\theta \right),$$

then rotating $\mathcal{F}'$ by $-\varphi$. Notice that this rotation normalization method fails when $\mathcal{F}'$ exhibits a central symmetry. However, unlike a classical rotation normalization computing the direction of the principal axis, it has the advantage of assigning the same weight to all points in $\mathcal{F}'$, and hence to be more robust to the noise affecting its boundary.

Putting all the steps together, the affine invariant normalization of a solid shape $\mathcal{F}$ is the set of points (x_N, y_N) given by

$$\begin{pmatrix} x_N \\ y_N \end{pmatrix} = \begin{pmatrix} \cos \varphi & \sin \varphi \\ -\sin \varphi & \cos \varphi \end{pmatrix} B_{\mathcal{F}}^{-1} \begin{pmatrix} x - \mu_{1,0} \\ y - \mu_{0,1} \end{pmatrix},$$

for all $(x, y) \in \mathcal{F}$.

As seen in Fig. 4.1, a classical problem of this kind of normalization is its lack of robustness. Too strong deformations lead to a bad estimation of the moments.

Fig. 4.1 Cohignac's normalization. Left column: original images. Right column: affine normalization using the moments. The middle and bottom original images were obtained from the top original image by a numerical affine transformation. Even in this ideal framework, the normalized solid shapes are not superimposable at all: the moment-based normalization is not robust. Compare with the local normalization proposed in the next section (the middle and bottom original images were deformed by the same transformation as in Fig. 4.3)

4.1.2 Application to the MSER Normalization Method

We have described in Sect. 2.2 how stable image extremal regions were extracted from an image by the MSER method, a variant of Monasse's Fast Level Set Transform. The MSER extraction is a first step to stereo baseline or object tracking algorithms in which high speed is required. The quick and affine invariant comparisons of MSERs taken from different images are performed by direct application of the affine normalization described in the above-section 4.1. Once MSERs are computed in two images to be compared, the affine covariance of MSER detection permits to compute affine invariant moments of these regions and to perform quick comparisons. The MSERs are normalized as explained in Sect. 4.1: the covariance matrix is diagonalized and then the linear transformation performing its diagonalization is applied to each region. As a consequence rotational invariants over the normalized region can be used to compare them. This procedure is affine invariant and yields potential candidates to a match. However, the final check in the original method [118] is made by using correlation. Invariant descriptions are only used as a preliminary test. The normalized circular regions are correlated (for all relative rotations). Thus the MSER procedure is an interesting variant of what has been described in Sect. 4.1. Yet the preceding section pointed out the lack of robustness of the global affine normalization by moments and the need for a more accurate normalization. This will be the object of the next section, where normalization is based on robust flat parts of shapes.

4.1.3 Geometric Global Normalization Methods

The geometric global normalization method described in continuation is based on robust directions given by the bitangent lines and the flat pieces of a solid shape boundary or level line $\mathcal{L}$. In the previous method, the second order moments of the moment based global normalization were used to find principal shape directions. The bitangent lines and flat parts will now play that role and lead to a much more reliable geometric normalization. A similarity invariant and an affine invariant global normalization methods are proposed here. The best way to describe such methods is to directly give a self-explanatory algorithm. In the following level lines are parameterized by length.

4.1.3.1 Similarity Invariant Normalization

For each shape $\mathcal{F}$ with boundary $\mathcal{L}$, and for all robust straight line $\mathcal{D}$ computed from $\mathcal{L}$:

1. Translate $\mathcal{F}$ so that its barycenter becomes the origin of the plane.
2. Scale $\mathcal{F}$ so that its boundary has unit length.
3. Rotate $\mathcal{F}$ with respect to the origin so that the robust direction is horizontal.
4. Define the starting point of the parameterization of $\mathcal{L}$ as the intersection with positive ordinate between the vertical axis and the boundary of the solid shape. In case of ambiguity, choose the closest one to the origin.

4.1.3.2 Affine Invariant Normalization (Positive Determinant)

The procedure is illustrated in Fig. 4.2. For each robust straight line $\mathcal{D}$ computed from $\mathcal{L}$:

1. Consider the straight line passing through the barycenter G of $\mathcal{F}$, which is parallel to $\mathcal{D}$. Consider the intersection between $\mathcal{F}$ and the half-plane defined by this straight line which does not contain $\mathcal{D}$; call G_1 its barycenter, and G_3 the barycenter of the complementary part of $\mathcal{F}$.
2. Now consider the straight line passing through G_1 and G_3. It splits the solid shape into two parts, let G_2 and G_4 be their barycenter, such that $(\overrightarrow{G_3G_1}, \overrightarrow{G_2G_4})$ is directly oriented. (The lines G_1G_3 and G_2G_4 intersect at G.)
3. Points $\{G, G_1, G_2\}$ define an affine basis. Normalize $\mathcal{F}$ by applying to it the affine transformation mapping $\{G, G_2, G_1\}$ into $\{(0,0), (1,0), (0,1)\}$.
4. Define the starting point of the parameterization of $\mathcal{L}$ as the intersection with positive ordinate between the vertical axis and the boundary of the normalized solid shape. In case of ambiguity, choose the closest one to the origin.

The proof of the next proposition is straightforward from the preceding algorithms.

Proposition 5. *Let $\mathcal{L}_1$ and $\mathcal{L}_2 = A\mathcal{L}_1$ be two curves such that $\mathcal{L}_1$ is deduced from $\mathcal{L}_2$ by a similarity (resp. affine) transformation, denoted by A. Then the sets of all normalized geometric curves obtained by the above normalization algorithms applied to all bitangent lines are identical.*

Proof. There is by A a one-to-one correspondence between the bitangent lines of $\mathcal{L}_1$ and $\mathcal{L}_2$ and the two above algorithms then describe a similarity (resp. affine) invariant procedure leading to identical normalized shapes. $\square$

The result was enounced for bitangent lines only, as the robust lines also obtainable from flat pieces of the curves are not *stricto sensu* similarity or affine invariant. Notice, however, that the use of flat zones is unavoidable to encode convex shapes, which have no bitangent lines. Moreover, under reasonable zoom factors, flat parts are preserved. Flat parts are often detected as tangent lines at inflexion points (which are conserved by affine transformations).

Figure 4.3 shows an example of global affine invariant normalization. The shapes are the same as in Fig. 4.1. Notice that the normalization is much more stable than in the moment-based approach.

4.2 Semi-Local Normalization and Encoding

The necessity of a local shape encoding has been emphasized enough. So the preceding sections on global encoding are mere essays towards a local one. This will be actually a simple adaptation.

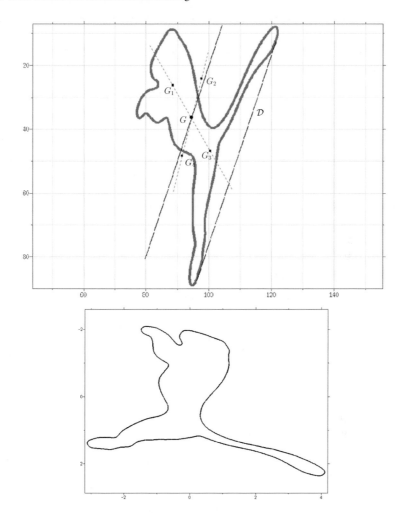

Fig. 4.2 Global affine invariant normalization based on the bitangent line $\mathcal{D}$. Top: definition of points G_1, G_2, G_3 and G_4. Bottom: the normalized solid shape

4.2.1 Similarity Invariant Normalization and Encoding Algorithm

Given a level line $\mathcal{L}$, for each flat piece or for each bitangent line do the following (this procedure is illustrated in Fig. 4.4):

a) Call P_1 the first tangency point and P_2 the other one (for flat pieces, P_1 and P_2 are the endpoints of the detected flat segment). Consider the tangent line $\mathcal{D}$ containing these points;

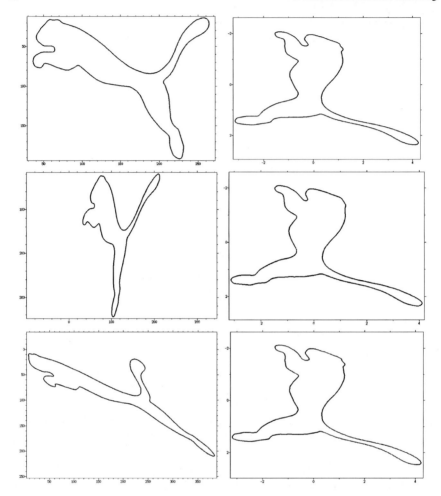

Fig. 4.3 Global affine invariant normalisation based on robust directions. Left column: boundaries of the original image (on top) and of two affine deformations of it (the same ones as in
Fig. 4.1). Right column: corresponding affine normalizations based on a bitangent line. The normalized shapes are very close; this is not the case with the invariant moment method

b) Call $\mathcal{P}_1$ the first tangent line to $\mathcal{L}$ which is orthogonal to $\mathcal{D}$, starting from P_1 in
   the negative direction. Call $\mathcal{P}_2$ the first tangent line to $\mathcal{L}$ which is orthogonal to
   $\mathcal{D}$, starting from P_2 in the positive direction.
c) Find the intersection points between $\mathcal{P}_1$ and $\mathcal{D}$, and between $\mathcal{P}_2$ and $\mathcal{D}$. Call
   them R_1 and R_2 respectively;
d) Store the *normalized* coordinates of N equi-distributed points over an arc on
   $\mathcal{L}$ of length $F \cdot \|R_1 R_2\|$, centered at C, the intersection point of $\mathcal{L}$ with the
   perpendicular bisector of $[R_1 R_2]$ (the first intersection starting from P_1). By

normalized coordinates one understands coordinates in the similarity invariant frame defined by points R_1, R_2 mapped to $(-\frac{1}{2}, 0), (\frac{1}{2}, 0)$ respectively.

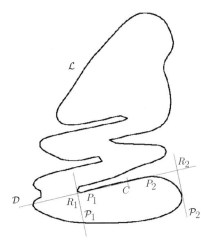

Fig. 4.4 Similarity invariant semi-local encoding based on a flat part of straight line $\mathcal{D}$

Two implementation parameters, F and N, are involved in this normalization procedure. The value of F determines the normalized length of the LLD. It has to be chosen keeping in mind the following trade-off: if F is too large, LLDs will be too long to deal with occlusions, while if it is too small, LLDs will not be discriminatory enough. The choice of F brings up a classic dilemma in shape analysis addressed in the bibliographical notes of this chapter (Sect. 4.3): locality *versus* globality of shape representations. The choice of N is less critical from the shape representation viewpoint since it is just a sampling precision parameter. Its choice results from a compromise between accuracy of the LLD and computational load.

Figure 4.5 shows some LLDs extracted from a single boundary, taking $F = 5$ and $N = 45$. Notice that the representation is quite redundant and yields LLDs describing the boundary over a wide range of scales. This redundancy increases the possibility of recognizing shapes subject to partial occlusions or other local perturbations.

All the experiments in Chap. 6 concerning matching based on this semi-local encoding (Sect. 6.1) were carried out using $F = 5$ and $N = 45$. These parameters can be fixed once for all, and they are not to be tuned by the user.

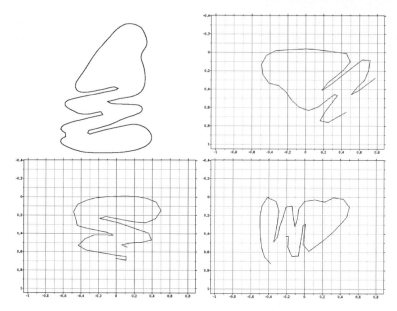

Fig. 4.5 Example of semi-local similarity invariant encoding. The line on the top-left generates 19 LLDs ($F = 5$, $N = 45$). Twelve of them are based on bitangent lines, the other ones are based on flat pieces. The representation is of course redundant. Three normalized LLDs, two deriving from bitangent lines, and one from a flat piece are displayed

4.2.2 Affine Invariant Normalization and Encoding Algorithm

The affine invariant representation of a level line $\mathcal{L}$ is computed by applying the following procedure for each flat piece or bitangent of $\mathcal{L}$ (this procedure is illustrated in Fig. 4.6):

a) Call P_1 the first tangency point and P_2 the other one (for flat pieces, P_1 and P_2 are the endpoints of the detected flat segment). Consider the tangent line $\mathcal{D}$ to these points;

b) Starting from P_2, find the next tangent to $\mathcal{L}$ which is parallel to $\mathcal{D}$. Call it $\mathcal{D}'$;

c) Consider the straight lines which are parallel to $\mathcal{D}$ and lay at $1/3$ and $2/3$ of distance from $\mathcal{D}$ to $\mathcal{D}'$. Call them $\mathcal{D}_1$ and $\mathcal{D}_2$ respectively;

d) Starting from P_2, find the next intersection points between $\mathcal{L}$ and $\mathcal{D}_1$, and $\mathcal{L}$ and $\mathcal{D}_2$. Consider the straight line $\mathcal{T}_1$ defined by these two points.

e) Starting from P_1, find the previous tangent to $\mathcal{L}$ parallel to $\mathcal{T}_1$, and call it $\mathcal{T}_2$;

f) Define points R_1, R_2, and R_3 as the intersections between $\mathcal{D}$ and $\mathcal{T}_2$, $\mathcal{D}$ and $\mathcal{T}_1$, and $\mathcal{D}'$ and $\mathcal{T}_2$ respectively;

g) Points R_1, R_2, R_3 define an affine basis . The affine normalization is fixed by mapping $\{R_1, R_2, R_3\}$ into $\{(0,0), (1,0), (0,1)\}$ if $\{R_1, R_2, R_3\}$ is a direct frame, and into $\{(0,0), (1,0), (0,-1)\}$ if not.

h) Encoding: consider the intersection point between $\mathcal{L}$ and the straight line equidistant from $\mathcal{D}$ and $\mathcal{D}'$ (the first one starting from P_2). Call it C. Normalize the portion of $\mathcal{L}$ having normalized length $F/2$ at both sides of C. Store N equidistributed points over the normalized piece of curve.

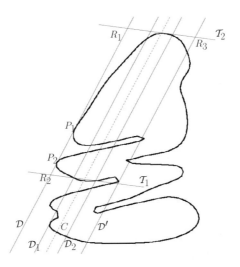

Fig. 4.6 Affine invariant semi-local encoding. The encoded LLD is based on the bitangent through P_1 and P_2

As for the similarity invariant normalization, implementation parameters were fixed once for all to $F = 5$ and $N = 45$. Figure 4.7 shows all LLDs extracted from a single boundary for this choice of parameters. Notice that the encoding is less redundant than for the similarity encoding procedure. This is because the construction of affine invariant local frames imposes more constraints on the curve than for similarity invariant frames.

4.2.3 Typical Number of LLDs in Images

The number of LLDs of a gray level image depends on the complexity of its level lines. Indeed, the number of LLDs is roughly proportional to the number of inflexion points. Textured images have in general many LLDs since their level lines are quite complex. To give an order of magnitude, the level lines of a database of 23 natural images of different type were encoded using the similarity encoding procedure described above. The level lines were respectively:

1. All meaningful boundaries;
2. Only maximal meaningful boundaries;
3. Maximal meaningful boundaries with local contrast;

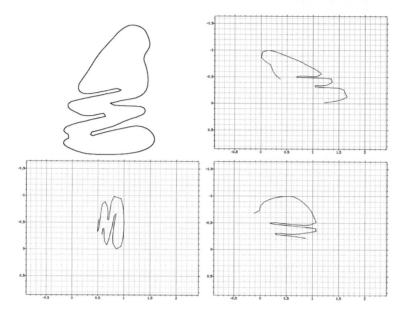

Fig. 4.7 Example of semi-local affine invariant encoding. The line on the top-left generates 7 LLDs ($F = 5$, $N = 45$); three of them are represented here

4. Cleaned (see Sect. 2.4.1.1) maximal meaningful boundaries with local contrast.

The number of LLDs per pixel and the CPU time of the encoding per pixel were measured. See Tab. 4.1.

Table 4.1 Number of LLDs encoded by the similarity semi-local encoding algorithm. Using local boundaries and the cleaning procedure makes the encoding much faster. In addition, the LLD dictionaries are shorter, but they experimentally contain all characteristic pieces of objects boundaries. The matching phase complexity is directly proportional to the number of pairs of LLDs, one taken in each image to be matched. Thus, this simplification is algorithmically quite fruitful

	#LLDs/pixel	CPU (s)/pixel
all MB	0.1458	0.0024
maximal MB	0.0528	0.0006
local MB	0.0310	0.0004
cleaned local MB	0.0132	0.0002

The typical time for encoding the meaningful boundaries is between 10s and 1min on a 1Ghz computer. The gain from meaningful boundaries to maximal meaningful boundaries is obvious and due to the elimination of redundancies in the level lines tree. The gain when using local meaningful boundaries is just empirical. Indeed, it is possible to construct images with more local meaningful boundaries than maximal meaningful boundaries. Since the cleaning procedure removes some parts of the level lines, the encoding is logically faster and the LLD dictionary shorter.

Remark 3. The semi-local encoding methods described above may not be local enough, particularly the semi-local affine invariant encoding. The normalized LLDs shown in Fig. 4.7 illustrate this problem. In fact, the main cause of non locality of the proposed semi-local normalization procedures is not the length of the encoded piece of curve (which could actually be controlled by parameter F), but the construction of the invariant frames (the lack of locality of this construction can be seen in Fig. 4.6). Normalization methods based on more local information are thus needed, in order to perform better in the presence of occlusions. Semi-local geometric normalization methods using area-based techniques, similar to the ones that we used for global encoding , could be explored.

4.3 Bibliographic Notes

The level lines extraction/smoothing/geometric encoding method described in this book was first introduced by Lisani *et al.* [108, 109]; the third geometric encoding stage described in this chapter is inspired from this reference and from Rothwell's work on invariant indexing [151]. The next subsections review and attempt to classify a wide number of antecedent shape encoding methods. Drawbacks will be pointed out without going into details (for example sensitivity to noise, occlusion, or deformation).

4.3.1 Geometric Invariance and Shape Recognition

We first review some references about the invariance issue in shape recognition. Shapes subject to weak perspective distortions, are easily recognized by humans. The geometric invariance requirement for shape recognition was already discussed in Chap. 1, Sect. 1.2.1. We claimed that in a general setting, affine invariance should be considered, while similarity invariance can suffice for a large class of particular applications. Such a claim was based on the following arguments and articles:

- Projective transformations are shown not to behave well with regard to shape matching, because they allow the mapping of a large class of curves to a curve which is arbitrarily close to a circle. (Thus, for example, a rabbit and a duck are almost projective equivalent [11, 12].)
- Despite some interesting attempts [63], there is no practical way to define projective invariant local smoothing. Thus affine invariant smoothing is the best possible [5].
- Since projective transformations are differentiable, they can be locally approximated by affine transformations (for which invariant smoothing is well defined),

and these approximations are particularly accurate under weak perspective distortion.

All in all affine distortions have to be thought of as local distortions. This is not really restrictive, since the locality of shape representation was already required in order to deal with occlusions and with the figure-background problem (see also Chap. 1, Sect. 1.2.1).

4.3.2 Global Features and Global Normalization

The simplest recognition methods are global in the sense that the extracted features are computed over the whole solid shape. Since they mix global and local information, they are sensitive to occlusions (part of the solid shape is hidden) or insertion (a part is added to the solid shape). This makes them inappropriate for general applications and restricts their use to a few specific applications where the observed objects do not overlap. The global features are in general scalar numbers computed over the whole solid shape. In the case of closed curves, Fourier descriptors [100, 104, 148, 185] or invariant moments [61, 133] (following Hu [86]) can be used. Affine invariant scalars for global shape representation can also be derived from wavelet coefficients [97, 162]. Using wavelets allows one to capture some local shape information, but not to the point of being able to deal with occlusion (the invariant scalars are computed by using coefficients from different scales). Another well-known moment related global method is the Sclaroff and Pentland [160] *modal matching*. In this method, a physical elastic model of the solid shapes is considered. Shapes are represented by their ordered set of eigenvalues for the elastic model. This method permits relatively realistic shape deformations where the thin parts of the shape can alter more than the bulk.

An original approach using size functions was proposed by Frosini *et al.* [66, 67]. Size functions can be seen as tools to get information about the topology of any graph. Applied to shape recognition, the size function theory leads to nearly invariant descriptors which can be well adapted to perceptual matching since they rely on structural information. Methods based on moments or Fourier descriptors, as well as size function methods, face the same problem. How to define the relative weights of each moment or size function in a shape comparison distance? This choice is in general arbitrary or based on *ad hoc* arguments. Robustness against noise is another aspect of this problem. Since high order moments (or high frequency *modes* for the modal matching method) represent details or fine information about the shape, they can be contaminated by noise and should not be considered. But up to what order should moments be considered?

Moments-based normalization methods such as the one presented in Sect. 4.1.1 have been extensively used in shape recognition. As noticed before, these methods suffer from two stability problems. First because of the dependence on second order moments the points in the contour of the shape strongly influence the result, making normalization quite sensitive to contour deformations. This effect can be reduced by considering robust norms such as Geman-McClure's ρ-function [69] for the estimation of the principal axis instead of the standard quadratic norm [171, 42]. Second, an error in the identification of the principal axis when the shape eigenvalues are close may yield completely different normalizations.

More stable global normalization methods can be built by considering bitangent lines, as in the geometric normalization method proposed in Sect. 4.1.3. In [143] affine invariant frames for global shape normalization are built by considering the pair of tangency points of the curve with the bitangent line and an extra point which can be the barycenter of the solid shape. This is not as stable as the geometric global normalization proposed here since the position of the bitangency points is not as robust as the direction of the bitangent line.

The scale-space representation of level lines can also be used to derive invariant representations. One such method can be found in Alvarez *et al.* [6], where shape invariants are based on the evolution of area and perimeter of the solid shapes surrounded by the level line undergoing the affine scale space. Let us describe the seminal work by Mokhtarian and Mackworth [132]. A shape (in fact a Jordan curve) is smoothed by curvature motion. At each scale, the smoothed curve is reparameterized by the normalized arc length, and the position of inflexion points (zero-crossings of the curvature) is tracked. If σ denotes the scale and s the corresponding normalized arc length, the proposed multiscale representation of the shape consists of the set of 2-tuples (s_i, σ_i), corresponding to the position and the scale at which two inflexion points meet and vanish. The corresponding binary image in the (s, σ) plane has been called the *Curvature Scale Space* and is a similarity invariant representation. It can also be robust to noise if one only considers the information given by the scale space for scales larger than an *ad hoc* or arbitrarily fixed threshold. At first sight, this method seems to be able to deal with occlusion since curvature is a local property of curves. This is not the case, however, since at each scale curves are reparameterized by the normalized arc length, and occlusions or insertions can drastically modify the positions of points (s_i, σ_i).

4.3.3 Local and Semi-Local Features

While global features are in general defined to be geometrically invariant up to rigid transformations, the local or semi-local features defined in the shape recognition literature can be invariant or not.

Commonly used non invariant features are, for instance, sets of edges [116, 117]. Groups of features are more informative than individual local features, and consequently enhance the matching stages: chained edges [183] or edgels [144] (an edge element with a direction) can be considered.

In order to achieve (geometrical) invariant recognition, non invariant features must be compared by means of strategies dealing with invariance, thus leading to time consuming algorithms. Non invariant features will not be further discussed.

Invariant local features may be computed directly on the image, or after the shape has been extracted. Features can be differential or integro-differential invariants at some special points (like corners [159]) or regions (*e.g.* coherent regions [21, 180]) of the image. The computation of differential invariants is quite unstable even after smoothing the image, since it involves high order derivatives.

Weiss [178] proposes local projective invariants requiring the computation of fourth order derivatives of the curves. This is of course out of range for contours of solid shapes derived from real images. Sato and Cipolla [156] propose semi-local quasi-invariants of curves, which do not need high order derivatives. Nevertheless, their affine quasi-invariants involve second order derivatives. This still is unrealistic for curves extracted from real images even after a smoothing step. Nowhere in this whole book will derivatives be involved in the shape recognition process, not even a first derivative (tangents are not used, only bitangents). Cohen *et al.* [39, 87] propose to approximate curves with B-Splines, leading to a compact representation. This interpolation appears to be robust to noise, and an adequate matching algorithm allows for dealing with occlusions. Although this method seems promising, it suffers from the interpolation in itself, which depends on the original sampling of the considered curve.

Most local recognition methods involve curvature extrema of the curves bounding the solid shapes. These points are not affine invariants of curves, but are certainly from the perceptual viewpoint the most salient points of shapes. This was already pointed out by Attneave in his 1954 paper [13]:

> *Information is concentrated along contours (i.e., regions where color changes abruptly), and is further concentrated at those points on a contour at which its direction changes most rapidly (i.e., at angles or peaks of curvature).* (See Fig. 4.8).

Cohignac *et al.* [41] propose a multiscale curvature representation for shape recognition by considering curvature extrema of surfaces derived from a shape with the affine morphological scale space. This leads, for each shape, to a set of points of interest in $\mathbb{R}^3$. In such local shape recognition methods, shapes are represented by a finite code, composed of the coordinates of curvature extrema points. Recognition can then be made local or semi-local by comparing the codes through the partial Hausdorff distance [88]. Two variations based on this general method leading respectively to a similarity invariant and to a translation-rotation invariant recognition methods can be found in [8, 68]. Similar approaches consist in using boundary points which are tangent to bitangent lines, instead of the curvature extrema [145].

Up to here, mainly local invariant features have been discussed. Since very local invariants such as differential invariants suffer from noise while global ones (*e.g.* moment invariants) suffer from occlusions a suitable trade-off can be the use of semi-local features.

Lamdan *et al.* [101], followed by Rothwell [151, 152], have proposed semi-local descriptors of shapes, invariant up to similarity or affine transformations. (Rothwell *et al.* also propose projective invariant representations.) These features are based on the description of pieces of non-convex curves lying between two bitangent points

Fig. 4.8 (From [13]) Curvature extrema concentrate a large amount of shape information. Quoting Attneave: Common objects may be represented with great economy, and fairly striking fidelity, by copying the points at which their contours change direction maximally, and then connecting these points appropriately with a straight edge

(*i.e.* points at which the same straight line is tangent to the curve). Such features are affine invariant and the use of bitangent lines ensures robustness to noise. Lisani *et al.* [108, 109] improved this bitangent method by associating, with each bitangent to each level line, a local coordinate system and defining a local affine or similarity normalized piece of curve. They also added to the representation similar local invariant descriptions based on tangent lines to the curve at inflexion points. This leads to a more complete representation of level lines.

Some recent methods of image analysis rely on invariant points of interest. These points are singularities of the image related to zero-crossings as in Lowe [114], or to Harris points [80]. By using locally computed affine invariant moments, these points can also be made affine invariant [121]. The purpose is merely to extract an invariant neighborhood of the image, independently of the shape they may contain. However, since interest points are usually located near relevant parts of shapes (see Fig. 4.8), some accurate semi statistical descriptors can be defined. For instance, the descriptors of [114] are local distributions of the gradient direction in some invariant neighborhoods of the points of interest, and are used in [163] for retrieving image parts in video sequences.

Part III
Recognizing Level Lines

Chapter 5
A Contrario Decision: the LLD Method

Abstract In this chapter we will try to answer the question "does that shape element look like this one?", and to measure the confidence level of this answer. This confidence level will be computed as the probability that two observed shapes match just by chance. This requires an *a contrario* or *background* model, which will be accurately computed from the shape database itself. The goal is to reach very high recognition confidence levels and therefore very small probabilities in the background model. How can we estimate very small probabilities? This cannot been done by simple counting. Indeed, the number of required samples grows as the inverse of the probability to be computed. There is, however, a classical way to circumvent this impossibility. It is enough to use independence. The probability of a very unlikely event can be estimated accurately provided it is a conjunction of independent events whose probabilities are larger, and therefore observable.

5.1 *A Contrario* Models

5.1.1 Shape Model or Background Model?

In what follows, it is always assumed that shape elements have been normalized as covered in Chap. 4. Consider a given query shape element S and a database B of N shape elements. Let us also assume that a distance or (dis-)similarity measure d between shape elements is defined. Assume that we found $S' \in B$ such that $d(S, S')$ is small. (One of the main purposes of the following discussion is to define what small does mean here.) The observed similarity of S and S' can have two explanations:

- $\mathcal{H}_0$: S' is near S only *by chance*. For instance because N is very large and there are similar shapes around every shape;
- $\mathcal{H}_1$: S' is near S because of a real similarity. For instance both come from two photographs of the same object.

F. Cao et al., *A Theory of Shape Identification*. Lecture Notes in Mathematics 1948.
© Springer-Verlag Berlin Heidelberg 2008

A full model for Hypothesis $\mathcal{H}_1$ is a model of all aspects of all objects we want to recognize. Accurately defining such a model would require large sets of observations. It must therefore be limited to very specific shape types like (e.g.) individual letters. It also requires learning algorithms. Thus, we do not consider it feasible to model $\mathcal{H}_1$ in a general shape recognition setting.

A model for $\mathcal{H}_0$ (namely the casual resemblance to $\mathcal{S}$) will be called *background model* and is more affordable. A way to construct it will be proposed in Sect. 5.2.1. The decision between $\mathcal{H}_0$ and $\mathcal{H}_1$ is taken by comparing the distance $d(\mathcal{S}, \mathcal{S}')$ with some predetermined value δ and deciding that $\mathcal{H}_1$ holds whenever $d(\mathcal{S}, \mathcal{S}') < \delta$. Otherwise, $\mathcal{H}_1$ is rejected and the alternative hypothesis $\mathcal{H}_0$ is accepted. The quality of a statistical test is measured by the probability of taking a wrong decision. Two kinds of errors are possible: reject $\mathcal{H}_1$ for an observation $\mathcal{S}'$ for which $\mathcal{H}_1$ is actually true (non-detection or type I error), and accept $\mathcal{H}_1$ for $\mathcal{S}'$ although $\mathcal{H}_1$ is false (false alarm, or type II error). A probability measure can be associated with each type of error. Thus we have:

- The *probability of non-detection* or *probability of a miss* (associated with type I error)

$$\text{PM}(\mathcal{S}, \delta) \equiv \Pr(d(\mathcal{S}, \Sigma) \geqslant \delta | \mathcal{H}_1);$$

- The *probability of false alarms* (associated with type II error)

$$\text{PFA}(\mathcal{S}, \delta) \equiv \Pr(d(\mathcal{S}, \Sigma) < \delta | \mathcal{H}_0), \qquad (5.1)$$

provided $\Pr(\cdot)$ is a probability measure defined on the set of shape elements. From now on our convention is to use Greek letters for random shape elements and Roman letters for observed values.

Note that the background model is given by $\Pr(\cdot | \mathcal{H}_0)$. It is clear that the lower PM and PFA, the better the test. Yet it is also clear that PM and PFA cannot be independently optimized. The usual problem is to find a trade-off between these two probabilities.

Widely used techniques such as the *Bayesian test* or the *Neyman-Pearson* test often amount to threshold the likelihood ratio of the observation under $\mathcal{H}_0$ and $\mathcal{H}_1$ [149]. However, the practical limits of this theoretical framework are obvious. They indeed require the knowledge of the likelihood of both the hypothesis $\mathcal{H}_1$ and the counter-hypothesis $\mathcal{H}_0$. This is generally unrealistic if the aim is to recognize an unspecified query shape element. A generative model is indeed needed for the query shape element $\mathcal{S}$ if the likelihood of each different shape element Σ under hypothesis $\mathcal{H}_1$ is to be computed. In the Bayesian approach, it is also required and generally not possible to accurately compute the probability of non-detection $\Pr(d(\mathcal{S}, \Sigma) \geqslant \delta | \mathcal{H}_1)$. This probability indeed relies on an observation model (noise, blur, projective distortion, etc.). Such a model is possible in particular applications where there are hypotheses on the shapes being sought. No such assumption is made in the present context, which aims at a full generality. If two images have shapes in common, these shapes appear in very few instances, and classical methods do not allow for the construction of models from these samples.

On the other hand, it will be easier to model the probability of false alarm $\mathrm{PFA}(\mathcal{S}, \delta)$. It is in fact possible to take a decision just based on the background probability model for $\mathcal{H}_0$. Sure detection simply requires that this probability is very small. Section 5.2 explains how to compute such small probabilities.

5.1.2 Detection Terminology

In presence of multiple testing, the fact that a probability is small has little meaning *per se*. What matters is the *number of false alarms*. We refer to the textbook [54] for a detailed analysis of this number in various geometric contexts and of its properties. Let N denote the number of shape elements in the database.

Definition 8. The *Number of False Alarms* of the shape element $\mathcal{S}$ at a distance δ is

$$\mathrm{NFA}(\mathcal{S}, \delta) \equiv N \cdot \mathrm{PFA}(\mathcal{S}, \delta), \tag{5.2}$$

where $\mathrm{PFA}(\mathcal{S}, \delta)$ is defined in (5.1).

The number of false alarms is the expected number of the shape elements in the database whose distance to $\mathcal{S}$ is below δ, when it is assumed that $\mathcal{B}$ obeys the background model.

Thus we will call $\mathrm{NFA}(\mathcal{S}, d(\mathcal{S}, \mathcal{S}'))$ the *number of false alarms between a query shape $\mathcal{S}$ and a database shape $\mathcal{S}'$*.

Definition 9. A shape element $\mathcal{S}'$ is an *ε-meaningful match* of the query shape element $\mathcal{S}$ if

$$\mathrm{NFA}(\mathcal{S}, d(\mathcal{S}, \mathcal{S}')) \leqslant \varepsilon. \tag{5.3}$$

Considering ε-meaningful matches as pertinent detections is an *a contrario* decision. The above definition is justified next.

Proposition 6. *Under the assumption that the database shape elements are identically distributed following the background model, the expectation of the number of ε-meaningful matches with $\mathcal{S}$ is less than ε.*

Proof. Let Σ_j $(1 \leqslant j \leqslant N)$ denote the shape elements in the database, and χ_j the indicator function of the event e_j: Σ_j is an ε-meaningful match of the query $\mathcal{S}$ (*i.e.* its value is 1 if Σ_j actually is an ε-meaningful match of $\mathcal{S}$, and 0 otherwise). Let $R = \sum_{j=1}^{N} \chi_j$ be the random variable representing the number of shape elements ε-meaningfully matching $\mathcal{S}$.

The key point is that the linearity of the expectation allows the computation of $\mathbb{E}_{\mathcal{H}_0}(R)$, the expectation of R in the background model. It is instead difficult or impossible to estimate the probability law of R (even under $\mathcal{H}_0$) because of the unknown dependencies between the events e_j. Linearity yields $\mathbb{E}_{\mathcal{H}_0}(R) = \sum_{j=1}^{N} \mathbb{E}_{\mathcal{H}_0}(\chi_j)$. By definition of χ_j,

$$\mathbb{E}_{\mathcal{H}_0}(\chi_j) = \Pr(\Sigma_j \text{ is an } \varepsilon\text{-meaningful match of } \mathcal{S}|\mathcal{H}_0).$$

By definition, Σ_j is an ε-meaningful match of $\mathcal{S}$ if

$$\Pr(\Sigma' \in \mathcal{B}, \, d(\mathcal{S}, \Sigma') < d(\mathcal{S}, \Sigma_j)|\mathcal{H}_0) \leqslant \frac{\varepsilon}{N}. \tag{5.4}$$

Notice that the probability on the left hand side in (5.4) is itself a random variable. The probability of this event is less than $\frac{\varepsilon}{N}$. Indeed, let us denote by X_j the random variable $d(\mathcal{S}, \Sigma_j)$, and F the repartition function (under $\mathcal{H}_0$) of $d(\mathcal{S}, \Sigma)$. Hence, F is also the repartition function of X_j and the event on the left hand side of (5.4) also reads $F(X_j) < \frac{\varepsilon}{N}$. Lemma 2 (page 21) then implies that

$$\Pr\left(F(X_j) < \frac{\varepsilon}{N}|\mathcal{H}_0\right) \leqslant \frac{\varepsilon}{N}.$$

This yields

$$\mathbb{E}_{\mathcal{H}_0}(R) \leqslant \sum_{j=1}^{N} \frac{\varepsilon}{N} = \varepsilon. \qquad \square$$

This methodology does not enable an *a priori* estimate of the number of ε-meaningful matches in a database of shape elements extracted from a natural image (i.e. whose shape elements are not likely generated by the background model). This number is an output of the method. The idea behind the definition is that if all shape elements in the database were generated by the background model, then Hypothesis $\mathcal{H}_1$ should never be accepted. In this case, all ε-meaningful detections should be considered false alarms. On average, there are less than ε detections.

The lower ε, the surer the ε-meaningful detections. Of course, the same claim is true when considering distances: the lower the distance threshold δ, the surer the corresponding matches. But considering the NFA quantifies this confidence level. Actually, by monotonicity, the equation

$$\delta^*\left(\frac{\varepsilon}{N}\right) \equiv \sup\{\delta > 0, \mathrm{PFA}(\mathcal{S}, \delta) \leqslant \varepsilon/N\}$$

suitably defines a positive real number. The proposition that follows is then straightforward.

Proposition 7. *A shape element $\mathcal{S}'$ is an ε-meaningful match of the query $\mathcal{S}$ if and only if $d(\mathcal{S}, \mathcal{S}') < \delta^*\left(\frac{\varepsilon}{N}\right)$.*

Thus, selecting ε-meaningful matches is equivalent to selecting shape elements $\mathcal{S}'$ such that $d(\mathcal{S}, \mathcal{S}') < \delta^*\left(\frac{\varepsilon}{N}\right)$. In practice, the method consists in fixing ε, and the value $\delta^*\left(\frac{\varepsilon}{N}\right)$ remains implicit. Moreover, computing the NFA does not need any shape model for $\mathcal{S}$.

Definition 10. Let $\mathcal{B}_1$ and $\mathcal{B}_2$ be two databases containing respectively N_1 and N_2 shape elements. The *Number of False Alarms* of a shape element $\mathcal{S}$ (belonging to $\mathcal{B}_1$) at a distance δ is

$$\text{NFA}(\mathcal{S}, \delta) = N_1 \cdot N_2 \cdot \text{PFA}(\mathcal{S}, \delta). \qquad (5.5)$$

This situation corresponds to experiments in Chap. 6 where the shape contents of pairs of images are compared. Prop. 6 then stays true, that is to say if $\mathcal{B}_2$ is generated by the background model, the expected number of ε-meaningful matches between $\mathcal{B}_1$ and $\mathcal{B}_2$ is less than ε.

5.2 The Background Model

The advantages of the *a contrario* decision framework compared to directly setting a distance threshold between shape elements are clear. Simply setting $\varepsilon = 1$ allows at most one false alarm among meaningful matches (1-meaningful matches will also be simply referred to as *meaningful matches*). In all experiments of the next chapter we will check that setting $\varepsilon = 10^{-1}$ eliminates false detections. The detection threshold ε can be set uniformly whatever the query shape element and the database are. While fixing $\varepsilon = 0.1$ will solve the detection problem, we still wish to take advantage of lower values of the number of false alarms to quantify the certainty of each match. Thus, our aim will be to compute NFAs, no matter how small they are.

Consider the following heuristic argument. Assume that the distribution of $d(\mathcal{S}, \Sigma)$ is learned by empirical frequencies on a set of N shape elements. Then the lowest non null observable probability is $1/N$. If $\mathcal{S}$ is now sought for in another database also containing N shape elements, then the lowest attainable number of false alarms is $N \cdot \frac{1}{N} = 1$. This means that even if two shape elements $\mathcal{S}$ and $\mathcal{S}'$ are almost identical, such an empirical estimate of the NFA cannot ensure that the match is not casual. Indeed, an NFA equal to 1 means that on average one of the shape elements in the database matches $\mathcal{S}$ just by chance. Lowe [112] commented in 1985 the very same aporia and the very same solution:

> Due to limits in the accuracy of image measurements (and possibly also the lack of precise relations in the natural world) the simple relations that have been described often fail to generate the very low probabilities of accidental occurrence that would make them strong sources of evidence for recognition. However, these useful unambiguous results can often arise as a result of combining tentatively-formed relations to create new compound relations that have much lower probabilities of accidental occurrence.

Definition 11. A shape *background model* $\mathcal{H}_0$ is a probability model on a set of shapes such that the following assumptions hold. Each shape element $\mathcal{S}$ can be represented by a set of K features $x_1(\mathcal{S}), \ldots, x_K(\mathcal{S})$, each of them belonging to a metric space (E_i, d_i) ($i \in \{1, \ldots, K\}$). Then the random variables $\Sigma \mapsto d_i(x_i(\mathcal{S}), x_i(\Sigma))$ ($i \in \{1, \ldots, K\}$) are mutually independent.

From the partial distances d_i, a complete, global distance should be defined, in order to apply the results of Sect. 5.1.1. A possible choice could be the product distance d defined by

$$d(\mathcal{S}, \mathcal{S}') = \max_{i \in \{1,\ldots,K\}} d_i(x_i(\mathcal{S}), x_i(\mathcal{S}')). \qquad (5.6)$$

Nevertheless, there is no reason why the d_i should have the same order of magnitude. Instead, denote by $P_i(\mathcal{S}, \delta)$ the marginal probability

$$P_i(\mathcal{S}, \delta) = \Pr(d_i(x_i(\mathcal{S}), x_i(\Sigma)) \leqslant \delta | \mathcal{H}_0). \tag{5.7}$$

Let us define

$$\delta_i(\mathcal{S}, \mathcal{S}') = P_i(\mathcal{S}, d_i(\mathcal{S}, \mathcal{S}')) \tag{5.8}$$

i.e.

$$\delta_i(\mathcal{S}, \mathcal{S}') = \Pr(d_i(x_i(\mathcal{S}), x_i(\Sigma)) \leqslant d_i(x_i(\mathcal{S}), x_i(\mathcal{S}')) | \mathcal{H}_0). \tag{5.9}$$

We can also define the product distance

$$d(\mathcal{S}, \mathcal{S}') = \left(\max_{i \in \{1, \dots, K\}} \delta_i(\mathcal{S}, \mathcal{S}') \right)^K. \tag{5.10}$$

Despite denomination, this function is not necessarily a distance. However, $d(\mathcal{S}, \mathcal{S}')$ is small when observing random values $d_i(x_i(\mathcal{S}), x_i(\Sigma))$ smaller than $d_i(x_i(\mathcal{S}), x_i(\mathcal{S}'))$ occurs with a low probability. Hence d is a measure of dissimilarity which is relative to $\mathcal{S}$.

The purpose of this operation is the following: If Σ is a random shape element, $\delta_i(\mathcal{S}, \Sigma)$ also is a random variable. If the distributions of the distances d_i are given by densities, then δ_i is uniform in $(0, 1)$ by Lem. 2 (p. 21), whatever the law of Σ. Of course, the δ_i are independent if the d_i are independent, which is assumed in the background model.

The NFA between $\mathcal{S}$ and $\mathcal{S}'$ is still defined by

$$\mathrm{NFA}(\mathcal{S}, \mathcal{S}') = N \cdot d(\mathcal{S}, \mathcal{S}').$$

The next result immediately generalizes Prop. 6.

Corollary 1. *The expected number of ε-meaningful matches in a database of N shape elements generated by the background model is less than ε.*

Proof. The sketch of the proof follows the one of Prop. 6. By linearity of the expectation, it suffices to prove $\Pr(\mathrm{NFA}(\mathcal{S}, \Sigma) < \varepsilon | \mathcal{H}_0) < \frac{\varepsilon}{N}$. By definition, $\mathrm{NFA}(\mathcal{S}, \Sigma) < \varepsilon$ if and only if, for all $i \in \{1, \dots, K\}$,

$$\delta_i(\mathcal{S}, \Sigma) = P_i(\mathcal{S}, d_i(\mathcal{S}, \Sigma)) < \left(\frac{\varepsilon}{N} \right)^{1/K}.$$

By the independence assumption,

$$\Pr(\mathrm{NFA}(\mathcal{S}, \Sigma) < \varepsilon | \mathcal{H}_0) = \prod_{i=1}^{K} \Pr\left(P_i(\mathcal{S}, d_i(\mathcal{S}, \Sigma)) < \left(\frac{\varepsilon}{N} \right)^{1/K} | \mathcal{H}_0 \right).$$

But since P_i is exactly the repartition function of $d_i(\mathcal{S}, \Sigma)$, Lem. 2, (p. 21) applies and each probability on the right hand product is less than $\left(\frac{\varepsilon}{N} \right)^{1/K}$.

Hence,

$$\Pr(\text{NFA}(\mathcal{S}, \Sigma) < \varepsilon | \mathcal{H}_0) \leqslant \frac{\varepsilon}{N}. \qquad \square$$

5.2.1 Deriving Statistically Independent Features from Level Lines

The $P_i(\mathcal{S}, \delta)$ will be empirically learned on a size N database. The smallest attainable number of false alarms in a background model is thus of order $N \cdot \frac{1}{N^K} = N^{1-K}$. The number of features K should be large enough, so as to attain very small NFAs. But, it cannot be arbitrary large either. Indeed, digital images contain a finite amount of information. Therefore level line descriptors cannot be described by an infinite set of independent features. On the other hand, the features $x_i(\mathcal{S})$ should characterize a shape $\mathcal{S}$, so that $d(\mathcal{S}, \mathcal{S}')$ is small if and only if $\mathcal{S}$ and $\mathcal{S}'$ are similar. Finding a suitable trade-off between independence and completeness of the features is necessary.

The decision framework described so far is actually completely general in the sense that it can be applied to find correspondences between any kind of objects for which K statistically independent features can be extracted. We now concentrate on the problem of extracting independent features from level line descriptors (LLD). In order to make the shape recognition task reliable, shape features have to meet the three following requirements.

1) Completeness: Two LLDs are alike if and only if their features are alike;
2) Statistical mutual independence (more precisely, distances between features are independent);
3) Their number is as large as possible.

The first requirement means that the features describe shapes well. The second one is imposed in order to design the background model, and the third requirement is needed in order to reach low numbers of false alarms. The existence of a background model as defined in Def. 11 is not obvious. In particular, proving independence is not easy. The remainder of this section describes a possible construction of LLDs' features.

5.2.1.1 Semi-Local Encoding

First consider the semi-local encoding algorithm described in Chap. 4. Recall that an LLD is a piece of Jordan curve normalized in a local frame built on a bitangent or on a flat part. The construction to be described now yields a good trade-off achieving simultaneously the three feature requirements (see Fig. 5.1 for an illustration). Each normalized representation C is split into five pieces of equal length. Each one of these pieces is normalized by mapping the chord between its first and last points onto the horizontal axis, the first point being at the origin. The resulting normalized chunks are five features C_1, C_2, ..., C_5. These features ought to be independent;

nevertheless, $C_1, \ldots, C_5$ being given, it is impossible to reconstruct the LLD they come from. For the sake of completeness a sixth global feature C_6 is therefore made up of the endpoints of the five previous pieces in the normalized frame. For each LLD, the shape features introduced in Sect. 5.2 are made of the six shape codes $C_1, \ldots, C_6$. Using the notations introduced in the previous sections, $x_i(\mathcal{S}) = C_i, i \in \{1, \ldots, 6\}$; the distances d_i between them are L^∞-distances between corresponding pieces, parameterized by length. If $C_i(s)$ is such a parameterization, we will simply set $d_i(C_i, \tilde{C}_i) = \sup_s \|C_i(s) - \tilde{C}_i(s)\|$.

The independence hypothesis amounts to say that for shapes of the reference database each chunk of an LLD does not influence the other ones, and that scales do not interfere. It cannot be proved that this description provides a background model in the sense of Def. 11. However, its consistency with the theory of Sect. 5.1 will be empirically checked.

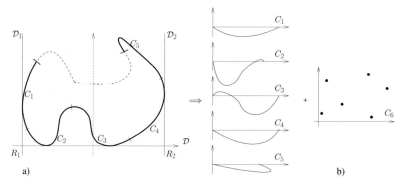

Fig. 5.1 Semi-local encoding procedure. Example of a similarity-invariant encoding. Sketch (a): original LLD in a normalized frame based on a bitangent line. Both ends of the LLD of length proportional to $\|R_1 R_2\|$, are marked with bold lines: this representation is split into five pieces C_1, C_2, C_3, C_4, and C_5. Sketch (b): each of them is normalized and a sixth feature C_6 made of the endpoints of these pieces is also built

Let us give some realistic orders of magnitude. In typical 512×512 images, the experimental number of extracted LLDs is about 10^4. Thus, the smallest number of false alarms when matching LLDs between two images is

$$10^4 \cdot 10^4 \cdot \frac{1}{(10^4)^6} = 10^{-16}.$$

In practice, for similar shapes, numbers of false alarms as small as 10^{-10} will be observed.

Remark 4. All LLDs are sampled with a fixed number of points, independently of their lengths in the image. While this solution makes the computation of distances between normalized LLDs faster, precision problems may arise when considering long LLDs presenting strong oscillations. For these pieces of level lines, normalized LLDs may simply not be accurate enough, leading to false detections. Notice however that false matches involving such long LLDs always show NFAs close to 1 (see Fig. 6.19 in Chap. 6 for an example).

5.2.1.2 Global Encoding (MSER)

A global curve normalization was also proposed in Chap. 4. The *a contrario* decision strategy is still valid, considering these normalized curves as shape elements, and building the features in a similar way as for the semi-local encoding. Precisely speaking, each normalized MSER is split into five pieces. The starting point was defined in Chap. 4 as the nearest point to the barycenter intersecting the vertical line to the bottom with a positive ordinate. In the same way as for semi-local encoding, each one of these pieces is normalized by mapping the chord between its first and last points on the horizontal axis, the first point being sent to the origin. The resulting chunks are five features $C_1, C_2, \ldots, C_5$. For the sake of completeness, a sixth global feature C_6 is made of the endpoints of the five previous pieces. The features are made of $C_1, \ldots, C_6$. The distances d_i between them are again L^∞-distances between corresponding chunks, parameterized by length.

5.3 Testing the Background Model

A way to test the reliability of the background model would be to check that the shape chunks are statistically independent on a large and realistic database. Yet this independence can be and has been disproved by a χ^2 test. Thus our purpose must be less ambitious. What we really need is to validate the computation of the expected number of detections (the Number of False Alarms, Prop. 6) in the background model.

Since very small NFAs cannot be observed, the comparison between predicted and observed NFAs will be performed on a very large database and for values of the NFA ranging from 0.01 to 10,000.

A first experiment compares the number of detections and its prediction when the LLDs database and LLD query are both random walks with independent increments. In this case the background model must be true, since the considered LLDs perfectly fit the independence assumption. Table 5.1 shows with no surprise that the Number of False Alarms is very accurately predicted for various database sizes. The number of detections with a NFA lower than ε is of order ε. Modeling LLDs with random walks is not realistic. We shall now check what amount of dependence between shape chunks might come from two dependence factors. First the fact that they are level lines (which forbids self-crossings) and second the normalization.

Table 5.2 shows the number of detections versus the number of false alarms for databases made of pieces of level lines (*not* normalized: the LLDs are made out

Table 5.1 Random walks. Average number of detections (over 10 samples) vs ε. The experiments were made with databases of different size (N from 10,000 to 100,000 LLDs)

ε / N	0.01	0.1	1	10	100	1,000	10,000
100,000	0	0	2.3	15.2	122.2	1,075.5	9,872.2
50,000	0.2	0.3	1.5	11.9	106.1	1,001.1	9,789.5
10,000	0	0	1.2	12.5	108.4	985.0	—

of 45 consecutive points on pieces of level lines). The LLDs have no self-crossing. Once again the number of detections is accurately predicted. The number of matches with a NFA less than ε is again of order ε.

Table 5.2 Pieces of white noise level lines with *no* normalization. Average (over 10 samples) number of detections vs ε. The three rows correspond to the size of the various databases, respectively $N = 101,438$ LLDs, $50,681$ LLDs and $9,853$ LLDs

ε / N	0.01	0.1	1	10	100	1,000	10,000
101,438	0.1	0.1	1.7	13.8	95.3	942.5	9,789.4
50,681	0	0	1.2	10.3	90.5	955.1	9,859.3
9,853	0	0.1	0.9	9.5	94.3	973.1	—

Let us now consider databases made of (*normalized*) LLDs extracted from pieces of level lines in white noise images. Table 5.3 shows that the number of detections is still of the same order of magnitude as the number of false alarms ε. Yet, it is not as precisely predicted as in the former experiments. Roughly speaking, this means that the *dependence mostly comes from the normalization procedure*, and not from the non-self-intersection constraint. Nevertheless, the order of magnitude is still correct, and does not depend on the size of the database.

Thus, the experiments confirm that we can adopt the Number of False Alarms under Helmholtz principle. According to this principle a match is relevant if it cannot happen in a white noise image. Table 5.3 shows that matches with a NFA lower than 0.1 are unlikely in white noise images. Requiring a good confidence in the detected matches thus leads to consider 0.1-meaningful matches in realistic experiments (see Chap. 6, Sect. 6.1).

One could suspect that the slight dependence shown in the last experiment comes from the normalization of very small (and therefore smooth) level lines. Table 5.4 checks that this is not true. It compares numbers of detections and of false alarms for a database of normalized long pieces of level lines extracted from white noise images. The results are not better than in the preceding experiment.

Table 5.3 Normalized pieces of white noise level lines. Average (over 10 samples) number of detections vs ε on databases with respective size $N = 104, 722$, $N = 47, 033$ and $N = 10, 784$ LLDs

ε $\diagdown$ N	0.01	0.1	1	10	100	1,000	10,000	100,000
$104, 722$	0.3	1.5	6.5	31.5	173.9	1, 264.4	9, 803.1	99, 899.5
$47, 033$	0.1	0.3	3.7	20.2	125.4	976.3	9, 854.2	—
$10, 784$	0	0.2	2.6	14.8	107.6	973.3	—	—

Table 5.4 Normalized long (more than 135 pixels) pieces of white noise level lines. Average (over 10 samples) number of detections vs ε on databases with respective size $N = 101, 743$, $N = 51, 785$ and $N = 11, 837$ LLDs

ε $\diagdown$ N	0.01	0.1	1	10	100	1, 000	10, 000	100, 000
$101, 743$	0	0.4	2.8	18.5	124.3	1, 123.2	9, 693.8	99, 921.0
$51, 785$	0	0.3	2.9	16.0	118.6	983.4	9, 800.4	—
$11, 837$	0	0.2	1.4	12.3	105.9	975.2	9, 974.7	—

5.4 Bibliographic Notes

5.4.1 Shape Distances

The shape matching problem is strongly related to the definition of adequate distances. The most commonly used distances are L^p distance, Mahalanobis distance [60, 165], Hausdorff distance [88], or Fréchet distance [4]. Miller, Younes and Trouvé [123, 124] (see also the more recent [20]) study the orbit of shapes via the action of diffeomorphic transformations, allowing in this way non-rigid transformations. Each transformation has a cost, and the distance between two shapes is the cost of the transformation with least energy between them. Similarity distance defined as the cost of an elastic deformation has been elaborated by [18]. Most of these distances are global and sensitive to local occlusions. However, they can be suitably modified to fit the locality requirement leading for instance to partial Hausdorff distance [88, 173, 145]. We refer the reader to general surveys by Alt *et al.* [3], Veltkamp *et al.* [173, 172], Loncarnic [111] and Dryden [57]. A review of more applied methods involved in Content-Based Image Retrieval (CBIR) systems is found in [174, 175].

Some global features allow shapes to be matched on other criteria than invariance with respect to a projective subgroup. For instance, a lot of work has been done on methods for matching shapes by minimizing the deformation energy involved in aligning one shape with another. One such method is modal matching [160], which takes a certain physical plausibility of the deformations into account, and thus accepts a larger class of invariance than geometric groups. Methods minimizing non-rigid energy deformations can also be based on local features, but they do not allow partial matching since all features are involved in the deformation energy. As an example Belongie *et al.* [22] propose to estimate the transformation leading from one shape to another when each shape is described by some points with a shape context (information about the points vicinity). Lisani *et al.* [108, 109] first defined shape elements as pieces of level lines. The normalization used in this book is basically the same. However, in Lisani's work, distance thresholds were chosen manually based on empirical testing.

5.4.2 A Contrario *Methods*

This chapter is mainly based on the paper [140] which introduced the *a contrario* method to match elements of level lines. *A contrario* detection frameworks are classical in the signal processing field, where a precise model of noise is often available. See for example an application to the detection of gravitational burst in Arnaud *et al.* [9], and another for the detection of small targets in cluttered environment in Chapple *et al.* [38]. In both cases in the absence of signal the data distribution is assumed to be a zero mean Gaussian with known variance.

An example of target detection in non-Gaussian images can be found in Watson and Watson [177]. The authors model the background of the considered images with a fractal model based on a wavelet analysis. Targets are detected as rare events with regard to this model.

The *a contrario* detection framework has recently been applied by Desolneux *et al.* for the detection of alignments [50] or contrasted edges [51], by Almansa *et al.* for the detection of vanishing points [2], by Stival and Moisan for stereo images [128], by Gousseau for the comparison of image composition [74] and by Cao for the detection of good continuations [31].

Another possibility that was investigated is to use the principal component analysis (PCA) [141]. Although PCA does not provide independent features but uncorrelated ones, the approximation does not seem to be critical. However, the completeness requirement (for the same number of features) is not satisfied with PCA. Moreover, shape elements do not form a vector space. The same remark holds for independent component analysis (ICA) [90], which assumes that the signals (here, shape elements) are linear mixtures of independent features.

Chapter 6
Meaningful Matches: Experiments on LLD and MSER

Abstract This chapter tests the shape matching method described in the previous chapter. Section 6.1 deals with the semi-local invariant recognition method. Both similarity and affine methods are considered, and a comparative study based on examples is presented. When images differ by a similarity, affine matching usually returns less matches because affine encoding is more demanding. Nevertheless, affine encoding proves more robust as soon as there is a slight perspective effect, and yields much smaller NFAs. We will also test an improved MSER method (namely a global affine matching algorithm of closed level lines). This algorithm works but we will point out a problem with convex shapes, which turn out to be very hard to distinguish up to an affine transformation. Finally the context-dependence of recognition will be illustrated by striking experiments on character recognition.

Now comes the time to check the applicability of the shape comparison scheme described in the previous chapters. All the experiments presented thereafter follow the same procedure: detection of meaningful boundaries (Chap. 2), affine invariant smoothing (Chap. 3, Sect. 3.3), similarity or affine normalization-encoding (Chap. 3 and 4), and then matching (Chap. 5).

6.1 Semi-Local Meaningful Matches

This section presents several experiments that illustrate all stages of the semi-local invariant recognition method, in particular the semi-local normalization procedures (Chap. 4) and the decision method (Chap. 5). Both similarity and affine versions will be compared.

F. Cao et al., *A Theory of Shape Identification*. Lecture Notes in Mathematics 1948.   93
© Springer-Verlag Berlin Heidelberg 2008

6.1.1 A Toy Example

This first experiment compares the performance of the affine invariant and the similarity invariant recognition methods on simple synthetic images. The role of such toy examples is to illustrate all the stages of the recognition methods. Figure 6.1 shows two synthetic images. LLDs from the image on the left (the *query* image) are sought in the right image (the *scene* image).

In the scene image, an affine distorted version of the symbol in the query image is included. The affine and the similarity semi-local invariant encoding algorithms, described in Chap. 4, were applied to the smoothed extracted boundaries before meaningful matches were detected in both cases.

(a)

(b)

Fig. 6.1 Toy example. (a) Original images. The image on the right contains an affine distorted version of the symbol in the left image. (b) Corresponding maximal meaningful boundaries

Using the semi-local affine invariant recognition method 44 LLDs were extracted from the query image's meaningful boundaries. These LLDs are represented by affine normalized codes of 45 points, as explained in Chap. 4. The same encoding procedure applied to the scene image led to 105 LLDs. Meaningful matches between these two sets of LLDs were detected. Following the rationale for the meaningful-ness computation presented in Chap. 5, a perfect match between LLDs would have reached a NFA of $44 \times 105/105^6 = 3.45 \, 10^{-9}$ (when the empirical distributions of distances to query LLDs are learned using only the considered scene image, as done here). But perfect matching is impossible, even with synthetic images. Indeed, the interpolation involved in the affine transformation of the image leads to boundaries that are not exactly the transformed boundaries of the original image. Another rea-son is as pointed out in Chap. 4 that flat pieces are not affine invariant (they are not even similarity invariant), and their position may vary.

This is exactly what can be observed in the experiment. All 42 detected mean-ingful matches between LLDs for the affine invariant framework (NFA < 1) are shown (superimposed) in Fig. 6.2(a). No false match was detected. The best match has NFA $= 5.4 \, 10^{-7}$ and the worst one $9.6 \, 10^{-1}$. These two matches are displayed in Fig. 6.3(a). The leftmost and middle images correspond respectively to the query and the scene LLDs, and the rightmost image shows their LLDs in the normalized frame, superimposed. The LLDs matching at NFA $= 9.6 \, 10^{-1}$ do not correspond exactly to the same piece of curve but they are still detected since they are close enough. This kind of instability is not really a problem since in general the encod-ing is redundant enough to capture better matches involving the same portions of the curve. This is illustrated in Fig. 6.2(b) where almost all the same pieces of boundary shown in Fig. 6.2(a) are still present with a meaningfulness $\varepsilon < 10^{-2}$.

Finally, notice that one of the nested boundaries of the symbol does not have any matched LLD while the other (which is almost symmetric to it) does. The ex-planation is that in the scene image one of these nested boundaries has a flat piece and is therefore encoded. In the other one no flat piece is detected. This will not be a problem because such quasi-convex curves are also encoded in parallel by the global method presented in Chap. 4, Sect. 4.1.

The second part of this experiment applies the semi-local similarity invariant recognition method to the same query and scene images. The similarity invariant method is not expected to perform better than the affine invariant one, the com-mon LLDs in the query and the scene images being related to each other by an affine transformation. However, it is interesting to know if the semi-local similar-ity invariant method still is able to retrieve some matches. In this second part of the experiment the same stages as in the previous one were followed, except for the normalization/encoding procedure. The semi-local similarity invariant encoding method described in Chap. 4 is used. In the query image 80 LLDs were extracted from its meaningful boundaries and 127 for the scene image. Notice that the simi-larity invariant encoding is more redundant than the affine invariant encoding. The explanation is simple. As pointed out in Chap. 4 (Sect. 4.2) the construction of the affine invariant semi-local frames imposes more constraints on the curve than the similarity invariant one. (These affine semi-local frames are also more global than

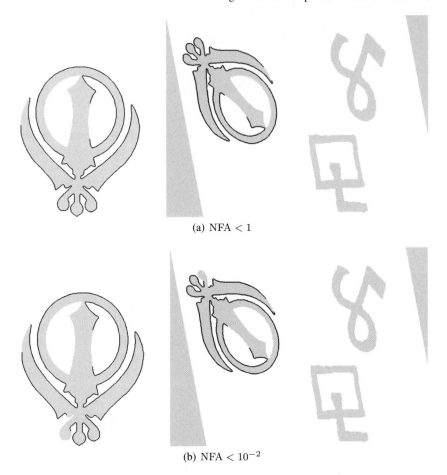

(a) NFA < 1

(b) NFA $< 10^{-2}$

Fig. 6.2 Affine invariant semi-local recognition. Meaningful matches (NFA < 1) between LLDs. No false match was detected

similarity semi-local frames, which makes them less robust to occlusion.) Perfect matches in this second part of the experiment could reach numbers of false alarms as low as $88 \times 127/127^6 = 2.66 \, 10^{-9}$. Here perfect matches cannot occur, mainly because boundaries are not related to each other by similarity transformations.

All 44 detected meaningful matches between LLDs (NFA < 1) for the similarity semi-local invariant recognition method are shown superimposed in Fig. 6.4. Figure 6.5 displays the matching LLDs in the images and in their corresponding normalized frame, for the largest and the lowest NFA ($2.5 \, 10^{-5}$ and $7.1 \, 10^{-1}$), as well as another example of matched LLD.

We see from the superimposed normalized LLDs that these LLDs are not as close as for the affine encoding. However, just look at LLDs in Fig. 6.5(a) and 6.5(c). Even though the query and the scene images are related by an affine transformation with

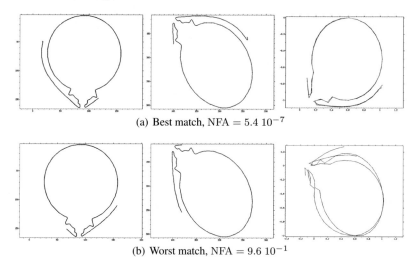

(a) Best match, NFA $= 5.4\ 10^{-7}$

(b) Worst match, NFA $= 9.6\ 10^{-1}$

Fig. 6.3 Affine invariant semi-local recognition. The matches showing the lowest and the largest NFA less than 1. Right column: both matched codes are superimposed. As expected, the first match is much more accurate

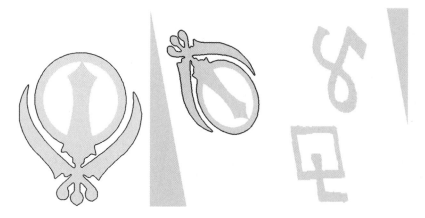

Fig. 6.4 Similarity invariant semi-local recognition. Meaningful matches (NFA < 1) between LLDs. No false match was detected

considerable shear and tilt, almost the entire shape is recognized with a high enough degree of confidence. The only exception is for the nested boundaries, which are too convex to be encoded by the semi-local method.

Part of the discussion presented in this section can be summarized in Fig. 6.6. The list of meaningful matches is ordered from best (lowest NFA) to worst (largest NFA), and the index i of this sorted list is plotted versus $-\log_{10}(\text{NFA}_i)$ where NFA_i is the NFA of the i-th best match. Such a function is plotted for the similarity and for the affine matches. The affine semi-local invariant matches reach lower

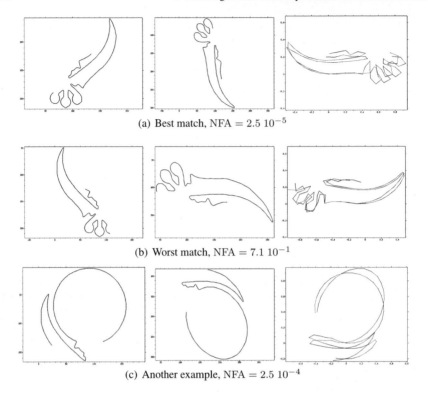

(a) Best match, NFA $= 2.5 \ 10^{-5}$

(b) Worst match, NFA $= 7.1 \ 10^{-1}$

(c) Another example, NFA $= 2.5 \ 10^{-4}$

Fig. 6.5 Similarity invariant semi-local recognition. The matches showing the lowest and the largest NFA

NFA. Notice that in both affine and similarity invariant recognition methods there are several matches that show small NFA, leading to the sure detection of common shapes.

6.1.2 Perspective Distortion

The affine method performs obviously better than the similarity method when dealing with images related through an affine transformation and not suffering from occlusion. This second experiment shows that the affine method also performs better than the similarity method when applied to real images related through moderately weak perspective transformations. The two images considered in this experiment (which we call Hitchcock experiment) are shown in Fig. 6.7 with their corresponding level lines. The resolution of these images is 640×480, which is enough to ensure good accuracy in the extracted level lines.

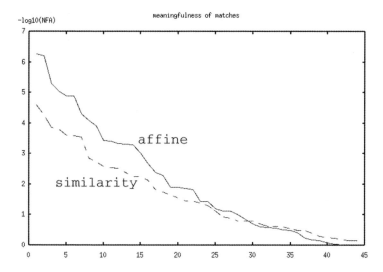

Fig. 6.6 NFA of affine and similarity semi-local invariant matches for the toy example. Both lists of meaningful matches are ordered from best (lowest NFA) to worst (largest NFA), and for each list, the index i of the sorted list is plotted *versus* $-\log_{10}(\text{NFA}_i)$, where NFA_i is the NFA of the i-th best match

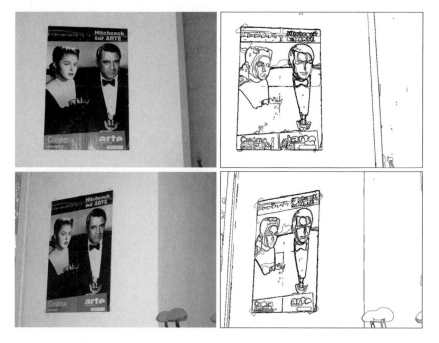

Fig. 6.7 Hitchcock experiment: original images and their corresponding level lines. Top: query image, 307 maximal boundaries were detected. Bottom: scene image, 266 maximal boundaries

For the affine semi-local invariant method, 1,150 and 853 LLDs were extracted from the query image and from the scene image respectively. The number of 1-meaningful matches detected was 517. In order to reduce the redundancy of the output, a greedy algorithm eliminates matched LLDs which share a large piece of curve with other LLDs presenting lower NFA. More precisely, if a pair of LLDs $(\mathcal{S}_1, \mathcal{S}_1')$ is an ε_1-meaningful match, and there exists another pair $(\mathcal{S}_2, \mathcal{S}_2')$ matching ε_2-meaningfully, with $\varepsilon_2 < \varepsilon_1$, such that $\mathcal{S}_1$ shares at least half of its length with $\mathcal{S}_2$, and if the same property holds for $\mathcal{S}_1'$ and $\mathcal{S}_2'$, then the pair $(\mathcal{S}_1, \mathcal{S}_1')$ is eliminated from the output list of matches. By this elimination of redundant matches, the list of meaningful matches is drastically reduced from 517 to 16 elements. This also shows how redundant the encoding is. These 16 matched LLDs are shown superimposed in Fig. 6.8. No false matches were detected, and all matches have their NFA below 0.1. The best match, shown in Fig. 6.9, reaches NFA $= 6.5 \, 10^{-11}$. This value is remarkably low, considering that ideal perfect matches in this experiment would have a number of false alarms of $1150 \times 853 / 853^6 = 2.5 \, 10^{-12}$ (when the empirical distributions of distances to query shape LLDs are learned using only the considered scene image, as done here).

Fig. 6.8 Affine invariant semi-local recognition method: meaningful matches between LLDs. No false matches were detected, and all detections show an NFA below 0.1. The lowest NFA is $6.5 \, 10^{-11}$

Fig. 6.9 Affine invariant semi-local recognition method: the match showing the lowest NFA $(6.5 \, 10^{-11})$

Figure 6.10 displays the meaningful matches detected using the similarity semi-local invariant recognition method. In this case, 2,033 and 1,463 LLDs were extracted from the query image and from the scene image respectively. As noticed in the toy example, the similarity method allows the extraction of more LLDs than the affine method. A total number of 244 meaningful matches (NFA < 1) were detected, and 26 matches were left after applying the greedy algorithm. The meaningful matches for the similarity method are shown in Fig. 6.10. The lowest NFA reached with the similarity method is $3.8\,10^{-8}$, and corresponds to the LLDs and the normalized LLDs presented in Fig. 6.11. Figures 6.10(b) and 6.10(c) present respectively the LLDs matching at $\varepsilon < 0.1$, and those for which the NFA is between 0.1 and 1. Notice that none of the 10^{-1}-meaningful matches are false matches, and that the corresponding LLDs are in general much more local than the LLDs matching in Fig. 6.10(c). Indeed, the more global the LLDs, the less accurate the similarity approximation of the underlying transformation, which is in fact a projective transformation. Two false matches, for which the NFA is larger than 0.1, are seen in Fig. 6.10(c). Figure 6.12 shows the LLDs of these false matches as well as the superimposed normalized LLDs represented in the normalized frame.

We end the discussion on the Hitchcock experiment with a comparison between the NFA of the meaningful matches for the affine and the similarity semi-local invariant methods illustrated in Fig. 6.13. The principle is the same as for the toy example from Sect. 6.1.1. The list of meaningful matches is ordered from best (lowest NFA) to worst (largest NFA), and the index i of this sorted list is plotted *versus* $-\log_{10}(\mathrm{NFA}_i)$, where NFA_i is the NFA if the i-th best match. Such a function is plotted for the similarity and for the affine matches. The affine semi-local invariant matches reach lower NFA. Notice that in both affine and similarity invariant recognition methods, there are several matches that show small NFA, leading to sure detections of common shapes.

6.1.3 A More Difficult Problem

Both in the toy example and the in Hitchcock experiment query and scene images represented different views of the same planar elements. Corresponding shapes were accurately described by the meaningful boundaries, leading to the detection of several matching LLDs with high detection confidence. In this subsection a more difficult example is considered. It consists in finding common LLDs between the pair of images in Fig. 6.14. Although at first sight these two different posters for the movie *Casablanca* are very similar they present many differences that considerably affect the topographic map and consequently the set of maximal meaningful boundaries. For instance the actors' faces in the query image (the one on top in Fig. 6.14) come from a snapshot while the scene image is a drawing.

In this example only the similarity semi-local invariant method is considered. The number of LLDs that were extracted from the query and the scene images were 3,540 and 8,554 respectively. Figure 6.15 shows the 1-meaningful matches

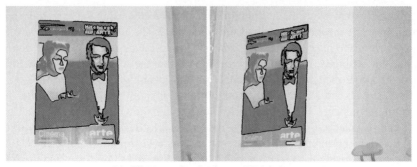

(a) All 26 matches having an NFA below 1

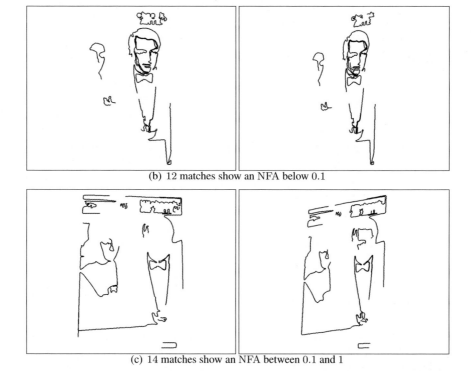

(b) 12 matches show an NFA below 0.1

(c) 14 matches show an NFA between 0.1 and 1

Fig. 6.10 Similarity invariant semi-local recognition method: meaningful matches between LLDs. Among the 26 matches having an NFA below 1, 12 are 10^{-1}-meaningful. Two false matches can be seen in (c); their NFA is above 0.1

Fig. 6.11 Similarity invariant semi-local recognition method: the match showing the lowest NFA ($3.8\,10^{-8}$)

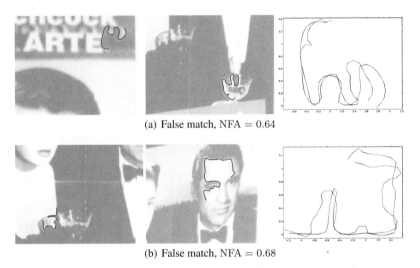

(a) False match, NFA $= 0.64$

(b) False match, NFA $= 0.68$

Fig. 6.12 Similarity semi-local invariant method: the two false matches. Their NFA are larger than 0.1

(*i.e.* matches for which NFA < 1) on the top row and the 10^{-1}-meaningful matches on the bottom row. The number of detected 1-meaningful matches was 211, which was reduced to 17 after applying the greedy algorithm. It seems that the majority of the relevant shape information that both images have in common has been detected. No meaningful match was found for the characters 'Casab', which are indeed quite different (up to a similarity) in both images.

Figure 6.16 shows the LLDs corresponding to the most meaningful match, for which NFA $= 1.1\,10^{-6}$. Such a low NFA is a consequence of the fact that this query LLD is so unusual that it is almost impossible that just by chance another LLD lies so close to it. A worthwhile remark here follows from the definition of the NFA given in Chap. 5. Suppose that two query LLDs $\mathcal{S}_1$ and $\mathcal{S}_2$ and two scene LLDs $\mathcal{S}_1'$ and $\mathcal{S}_2'$ satisfy $d(\mathcal{S}_1, \mathcal{S}_1') = d(\mathcal{S}_2, \mathcal{S}_2') = \delta$. Then if

$$\#\{\mathcal{S}' \in \mathcal{B} \text{ s.t. } d(\mathcal{S}_1, \mathcal{S}') \leqslant \delta\} < \#\{\mathcal{S}' \in \mathcal{B} \text{ s.t. } d(\mathcal{S}_2, \mathcal{S}') \leqslant \delta\},$$

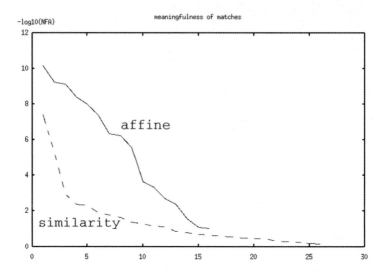

Fig. 6.13 Hitchcock experiment: NFA of affine and similarity semi-local invariant matches. Both lists of meaningful matches are ordered from best (lowest NFA) to worst (largest NFA), and for each list, the index i of the sorted list is plotted *versus* $-\log_{10}(\text{NFA}_i)$, where NFA_i is the NFA of the i-th best match

it follows that $\text{NFA}(\mathcal{S}_1, \mathcal{S}'_1) < \text{NFA}(\mathcal{S}_2, \mathcal{S}'_2)$. Hence for a given distance d the rarer a query LLD $\mathcal{S}$ (with respect to $\mathcal{B}$) the lower $\text{NFA}(\mathcal{S}, d)$. This makes sense. A rare LLD is more discriminatory than a banal one.

Figure 6.17 shows all the false matches detected at NFA < 1. They all have an NFA between 0.1 and 1. Finally, notice that all matches which semantically correspond to the same LLDs show NFAs below 0.1.

6.1.4 Slightly Meaningful Matches between Unrelated Images

The experiment presented in this subsection consists in looking for common LLDs between unrelated images. Two examples are considered. Query and scene images for the first experiment are shown in Fig. 6.18. All the matches for which NFA is below 1 are superimposed to the original images. 4,731 and 4,946 LLDs were extracted from the query and scene images respectively. Among all $4731 \times 4946 \approx 23 \, 10^6$ possible pairs of query-scene LLDs only 6 matches having NFA < 1 were detected. Their NFAs range from 0.21 to 0.97. The matched LLDs and their corresponding normalized LLDs are shown in Fig. 6.19. Numbers 1), 4) and 5), are simple (they are relatively short and do not present many oscillations) and match with pretty small distances. However, because of their banality they do not show lower NFAs. Matches number 2) and 6), while locally different, are quite similar at a coarse scale as can be seen from their superimposed normalized LLDs. For such long LLDs a

Fig. 6.14 Casablanca experiment. Left column: original images. Right column: corresponding level lines. The image on top was taken as query image

representation using 45 points may not be accurate enough. A finer sampling would probably have led to larger NFAs for that kind of matches.

A second example of LLDs common to unrelated images is shown in Fig. 6.20. The 22 LLDs extracted from the query image are searched in the 546 LLDs from the scene image on the left. The matched LLDs and their normalized LLDs are shown in Fig. 6.21. According to what was presented in Chap. 5, matches showing NFAs lower than 0.1 are not supposed to happen by chance (as would matches between LLDs extracted from random level lines). Thus some common reason must lay behind such an unexpected coincidence. In fact many shapes in images derive from natural or man-made objects having similar structures. In particular, many objects are built of parallel parts with equal length. This common feature was called *constant width* by Kanizsa.

6.1.5 Camera Blur

The last experiment deals with the semi-local invariant method for shape recognition. This experiment just aims at illustrating how the meaningful boundaries of

(a) NFA < 1

(b) NFA < 0.1

Fig. 6.15 Casablanca experiment. Meaningful matches between similarity invariant LLDs. Top: NFA < 1. Bottom: NFA < 10^{-1}. No false match can be seen

Fig. 6.16 The match with the lowest NFA. The query LLD (left image) matches the database LLD (right image). NFA = $1.1 \cdot 10^{-6}$

small objects are affected by the blur introduced when objects are far from the camera and how this problem can be solved by representing the query image at multiple scales. The same solution is applied in the celebrated SIFT method.

The query and scene images for this example are shown in Fig. 6.22, with their corresponding maximal meaningful boundaries. Images are displayed at the same scale. Figure 6.23 illustrates a detail of the maximal meaningful boundaries of the scene image corresponding to the region of interest for this experiment. Compare these boundaries with the ones extracted from the query image in Fig. 6.22 (on top

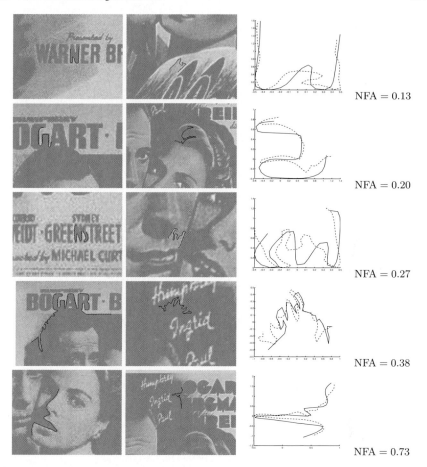

Fig. 6.17 The five false matches for NFA < 1 with their normalized LLDs. The leftmost and middle images correspond to the query and the scene LLDs respectively. The rightmost image shows their normalized LLDs superimposed. All false matches show NFAs between 0.1 and 1

right). The characters in the scene image have been almost completely destroyed and only a few similar LLDs can be observed.

Figure 6.24 on the top row shows the original image and two image reductions, by factors 4 and 8. The bottom row presents their corresponding maximal meaningful boundaries (followed by an affine shortening at scale $T = 0.5$, see Chap. 3, Sect. 3.3). Image reductions were performed using a prolate kernel.

Figure 6.25 shows the detected matches at NFA < 1 for each query image (the original image and the two reductions) with the scene image. When the LLDs of the original query image are searched only two matches having NFA < 1 are found (Fig. 6.25(a)). Both matches are correct and their NFAs are $8.8\,10^{-6}$ and $1.9\,10^{-4}$. Using as query image the image reduced by a factor 4 more meaningful matches are found and the best one reaches an NFA of $2.3\,10^{-10}$. In this case all matches

Fig. 6.18 Left: query image, 4,731 LLDs were extracted from this image. Right: scene image, the number of LLDs extracted from it was 4,946. Among the $23\,10^6$ pairs of query-scene LLDs, only six match with NFA < 1. The NFA of these matches range from 0.21 to 0.97

are correct (Fig. 6.25(b)). Finally using the query image reduced by a factor 8 even more meaningful matches are detected and reach lower NFAs. In this case the NFA of correct matches ranges from $2.1\,10^{-3}$ to $3.6\,10^{-12}$. A false match with NFA $= 7.6\,10^{-1}$ was detected but it corresponds to an artifact (a border effect) of the image reduction. See Fig. 6.25(c).

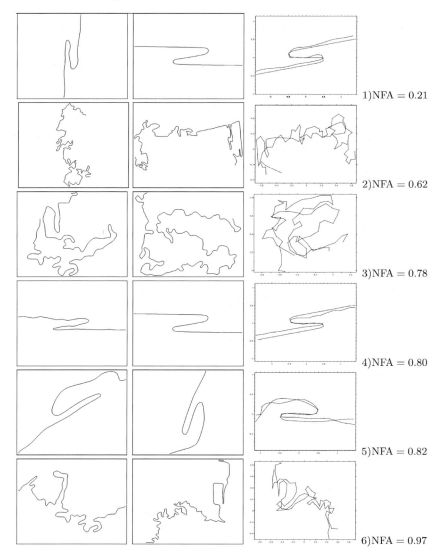

Fig. 6.19 The six false matches detected for NFA < 1 with their normalized LLDs. The leftmost and middle images correspond to the query and the scene LLDs respectively. The rightmost image shows their normalized LLDs superimposed. All false matches show NFAs between 0.1 and 1

Fig. 6.20 Puma experiment. Left: query image, from which 22 LLDs were extracted. Right: scene image; 546 LLDs were extracted from it. The two matches detected at NFA < 1 are superimposed to the original images

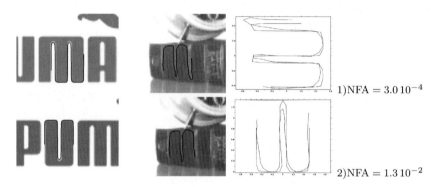

1)NFA $= 3.0\,10^{-4}$

2)NFA $= 1.3\,10^{-2}$

Fig. 6.21 Puma experiment: the two matches detected for NFA < 1, with their normalized LLDs. The leftmost and middle images correspond to the query and the scene LLDs respectively. The rightmost image shows their normalized LLDs superimposed. Such a conspicuous coincidence admits a better explanation than randomness: many shapes in images derive from natural or man-made objects having a common structure. For instance, many objects are built of parallel or equal-length parts

Fig. 6.22 Top row: query image and its maximal meaningful boundaries; 312 LLDs were extracted from this image. Bottom row: scene image and corresponding maximal meaningful boundaries. 1,859 LLDs were extracted from it

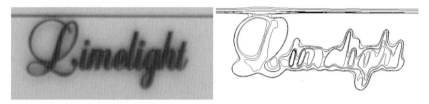

Fig. 6.23 Detail of the maximal meaningful boundaries of the scene image corresponding to the region of interest. The character boundaries have been very degraded by blur and smoothing

Fig. 6.24 Original query image and two image reductions. Left column: original image and corresponding maximal meaningful boundaries. Middle: image reduction by a factor 4 (324 LLDs were extracted from this image). Right: image reduction by a factor 8 (73 LLDs were extracted from this image). Reductions were performed using a prolate kernel

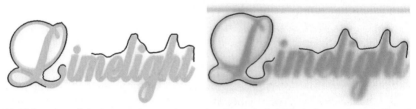

(a) Using the original query image. 2 matches have their NFA below 1 ($8.8\,10^{-6}$ and $1.9\,10^{-4}$)

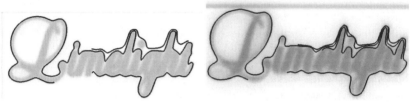

(b) Using an image reduction by 4 of the query image: 4 meaningful matches, with NFAs equal to $2.3\,10^{-10}$, $1.3\,10^{-6}$, $1.5\,10^{-5}$ and $5.7\,10^{-1}$

(c) Using an image reduction by 8 of the query image: 5 meaningful matches, at NFAs $3.6\,10^{-12}$ $7.4\,10^{-5}$, $4.6\,10^{-4}$, $2.1\,10^{-3}$ and $7.6\,10^{-1}$. The last one corresponds to a false match, but was introduced by an artifact in the image reduction procedure

Fig. 6.25 LLDs matched with the scene image using three different scales of a query image. The number of meaningful matches along with their meaningfulness increases when we consider image reductions. These image reductions simulate the effect of distance from the camera

6.2 Recognition Relative to Context

The recognition thresholds obviously increase with the rareness and decrease with the banality of the query LLD in the background database. In many experiments the background model was extracted from the image itself. With this choice the background model is the real *shape background*, which we can also call *shape context*. Choosing as background model the context of the shape or a neutral background made of arbitrary images will of course change the recognition thresholds. The aim of the experiment presented here is to show that the choice of the background can be steered by the user, depending on what he wishes. Assume the user wishes to pick all letters m in a scanned text. Then the question arises: Is the wish restricted to exactly the same lowercase m, with the same size, or are all lowercase m's welcome? Clearly the thresholds are not the same for both cases. They can actually be tuned by changing the background. To illustrate the concept four LLDs extracted from a sample character m (Fig. 6.26) were sought in 14 scanned pages by using the semi-local similarity invariant recognition method.

Two experiments were undertaken: in the first one the background model was built from the 14 scanned pages (79,376 LLDs) whereas in the second one the database came from 21 natural images (98,857 LLDs).

Figure 6.27 shows all matches in one of the 14 scanned pages with the first background model, which is basically the same text. Only m's with the same size and format as the query are recognized.

Figure 6.28 shows the recognitions with the generic database. Clearly the recognition thresholds were higher in the second case (Fig. 6.29). This result is fully coherent with the definition of *a contrario* recognition. In the first case, the focus is put on recognizing LLDs that share some common structure with a particular font m *among other fonts* against all other fonts. In the second case the focus is put on recognizing LLDs in the text image that share a common structure with m, *against arbitrary LLDs*. Thus, other similar characters will not be rejected. This explains why italic *m* were rejected in the first case and retrieved in the second.

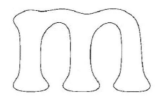

Fig. 6.26 Characters - the query curve

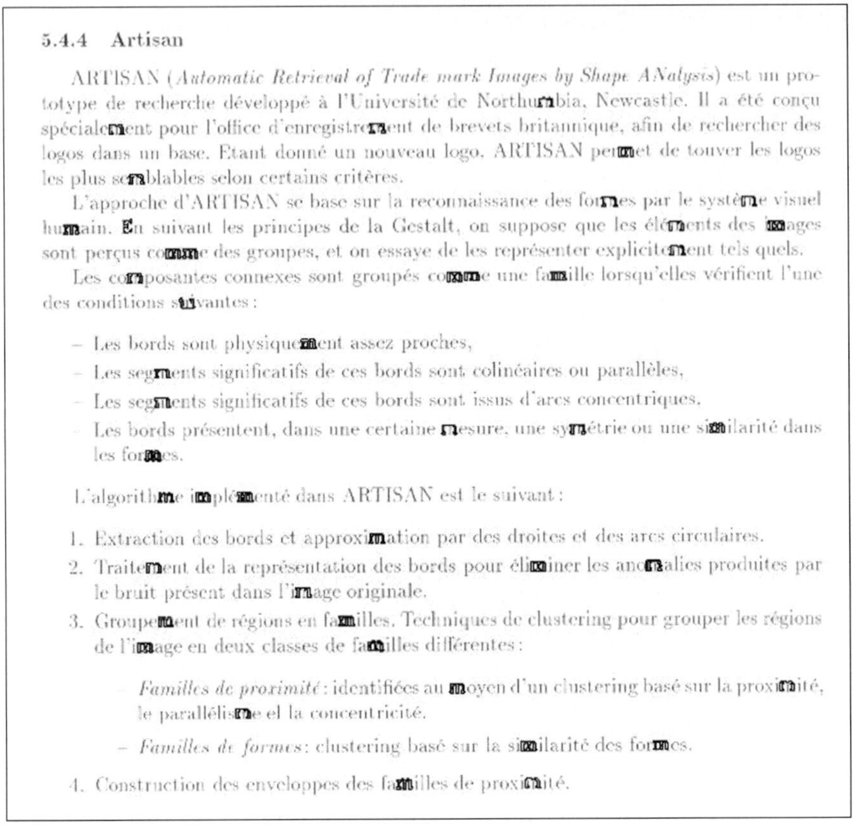

5.4.4   Artisan

ARTISAN (*Automatic Retrieval of Trade mark Images by Shape ANalysis*) est un prototype de recherche développé à l'Université de Northumbia, Newcastle. Il a été conçu spécialement pour l'office d'enregistrement de brevets britannique, afin de rechercher des logos dans un base. Etant donné un nouveau logo, ARTISAN permet de touver les logos les plus semblables selon certains critères.

L'approche d'ARTISAN se base sur la reconnaissance des formes par le système visuel humain. En suivant les principes de la Gestalt, on suppose que les éléments des images sont perçus comme des groupes, et on essaye de les représenter explicitement tels quels.

Les composantes connexes sont groupés comme une famille lorsqu'elles vérifient l'une des conditions suivantes :

– Les bords sont physiquement assez proches,
– Les segments significatifs de ces bords sont colinéaires ou parallèles,
– Les segments significatifs de ces bords sont issus d'arcs concentriques,
– Les bords présentent, dans une certaine mesure, une symétrie ou une similarité dans les formes.

L'algorithme implémenté dans ARTISAN est le suivant :

1. Extraction des bords et approximation par des droites et des arcs circulaires.
2. Traitement de la représentation des bords pour éliminer les anomalies produites par le bruit présent dans l'image originale.
3. Groupement de régions en familles. Techniques de clustering pour grouper les régions de l'image en deux classes de familles différentes :

   – *Familles de proximité* : identifiées au moyen d'un clustering basé sur la proximité, le parallélisme et la concentricité.
   – *Familles de formes* : clustering basé sur la similarité des formes.

4. Construction des enveloppes des familles de proximité.

Fig. 6.27 Character recognition when probabilities are estimated with a database of scanned text pages. A total number of 111 matches were detected. All m's having the same font as the query were retrieved. Only two matches with LLDs which do not belong to an m were found

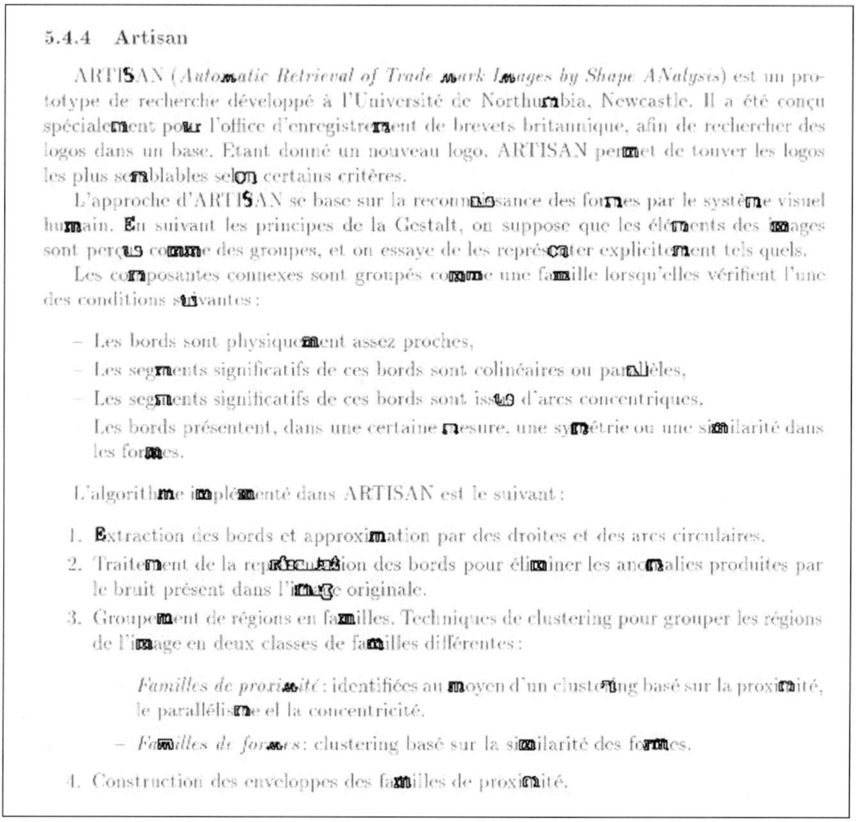

5.4.4 Artisan

ARTISAN (*Automatic Retrieval of Trade Mark Images by Shape ANalysis*) est un prototype de recherche développé à l'Université de Northumbia, Newcastle. Il a été conçu spécialement pour l'office d'enregistrement de brevets britannique, afin de rechercher des logos dans un base. Etant donné un nouveau logo, ARTISAN permet de trouver les logos les plus semblables selon certains critères.

L'approche d'ARTISAN se base sur la reconnaissance des formes par le système visuel humain. En suivant les principes de la Gestalt, on suppose que les éléments des images sont perçus comme des groupes, et on essaye de les représenter explicitement tels quels.

Les composantes connexes sont groupés comme une famille lorsqu'elles vérifient l'une des conditions suivantes :

– Les bords sont physiquement assez proches,
– Les segments significatifs de ces bords sont colinéaires ou parallèles,
– Les segments significatifs de ces bords sont issus d'arcs concentriques.
   Les bords présentent, dans une certaine mesure, une symétrie ou une similarité dans les formes.

L'algorithme implémenté dans ARTISAN est le suivant :

1. Extraction des bords et approximation par des droites et des arcs circulaires.
2. Traitement de la représentation des bords pour éliminer les anomalies produites par le bruit présent dans l'image originale.
3. Groupement de régions en familles. Techniques de clustering pour grouper les régions de l'image en deux classes de familles différentes :

   Familles de proximité : identifiées au moyen d'un clustering basé sur la proximité, le parallélisme et la concentricité.

   – *Familles de formes* : clustering basé sur la similarité des formes.

4. Construction des enveloppes des familles de proximité.

Fig. 6.28 Characters recognition when probabilities are estimated with a database extracted from natural images. 154 matches were detected. The corresponding distance thresholds obtained in this case were larger than those in Fig. 6.27

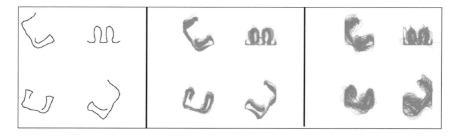

Fig. 6.29 Characters. Superimposition of the matched normalized LLDs. Left: the four query LLDs. Middle: all LLDs from the scanned text that match the corresponding query LLD superimposed; probabilities were estimated using the 14 scanned pages. Right: superimposed matched LLDs when probabilities were estimated with the database of natural images. The matching threshold is larger in the latter case

6.3 Testing *A Contrario* MSER (Global Normalization)

In Chap. 4 normalization methods invariant up to similarities or affine transformations were presented. This section shows several experiments on global shape matching that validate the normalization and the distance threshold derived from the number of false alarms (see Chap. 5). We deal here with closed level lines extracted as maximal meaningful in the level line tree. Thus the method tested in this section can be termed as an improved MSER method. It is improved in two ways: first because we use the geometric global normalization defined in Sect. 4.1.3 and second because we use the *a contrario* decision technique, which defines the right rejection thresholds. Before going to the experiments, recall that each level line leads to as many descriptors (or codes) as bitangent or flat parts in the curve.

6.3.1 Global Affine Invariant Recognition. A Toy Example

This first experiment illustrates the global recognition method with a simple example. The pair of images considered here involves the two images presented in Fig. 6.30, where the global meaningful matches have been superimposed. The image on the left was taken as query image, and the one on the right contains an affine distorted version of the query. MSERs were extracted by means of the global affine invariant normalization method (Chap. 4, Sect. 4.1.3), after extracting the meaningful boundaries and smoothing them with the affine curve shortening.

The detection of meaningful matches between all MSERs extracted from both images was performed using the detection method presented in Chap. 5. A single false match was detected with an NFA equal to 0.53. Matches between four different pairs of curves were detected. For each of these pairs, the best match (recall that between two curves, several matches between MSERs may exist) is shown in Fig. 6.31, with the corresponding NFA.

6.3.2 Comparing Similarity and Affine Invariant Global Recognition Methods

In this experiment we compare the performance of the affine and the similarity global recognition methods on two images whose maximal meaningful boundaries are shown in Fig. 6.32. The boundaries on the left were taken as query. The shapes in the scene images are strongly distorted, some by a projection on a cylinder (the bottle).

The aim was to find the character 'n' taken from the query image in the scene image.

Fig. 6.30 Toy example: original images with global meaningful matches superimposed. The image on the right contains an affine distorted version of the symbol in the left image

Figure 6.33 shows the detected 1-meaningful matches with MSERs extracted from the 'n' in the query image for the similarity invariant method. The query 'n' was represented by 10 MSERs. 36 matches were found in the scene image. The lowest NFA was 10^{-11}. Some false matches can be seen, but they all show NFAs between 0.7 and 1.

Figure 6.34 shows the matched MSERs when considering the global affine method. The query 'n' is still represented with 10 MSERs since it is the same 'n' that was considered for the global similarity matching (and there is always one MSER extracted for each bitangent line or flat part of the curve). A total of 35 matches showed an NFA below 1. The matches that actually correspond to the 'n' on the bottle, show NFAs that range from 10^{-15} to 10^{-8}. The NFA of matches which do not correspond to MSERs in the 'n' on the bottle, are between 10^{-3} and 1. However, some false matches are not really false but rather casual, since they correspond to other characters such as 'n' or 'u' that appear on the bottle (Minérale and Naturelle).

Figure 6.35 shows, for both methods, the matches showing the lowest NFA. The top row shows the normalized MSER for the global similarity invariant method and the bottom row shows the normalized MSER for the affine method. Notice that the pair of affine normalized MSERs are much closer to one another than the pair of similarity normalized MSERs. It seems then reasonable that the NFA reached with the global affine invariant method (10^{-15}) is lower than the one reached with the similarity method (10^{-8}).

In Fig. 6.36, the false match that shows the lowest NFA is presented for both methods. The top row shows the normalized MSER for the global similarity invariant method, and the bottom row shows the normalized MSER for the affine method. The NFA for the similarity invariant match was 0.7, and for the affine method $4.0\,10^{-3}$.

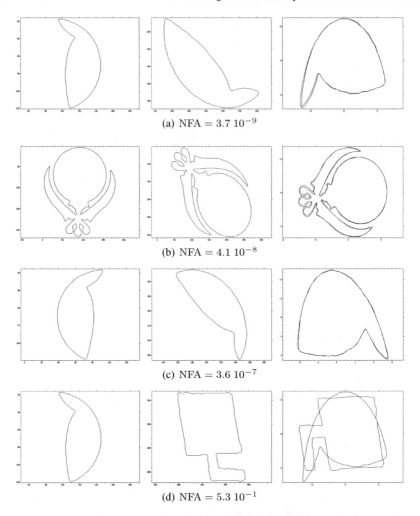

(a) NFA = 3.7 10^{-9}

(b) NFA = 4.1 10^{-8}

(c) NFA = 3.6 10^{-7}

(d) NFA = 5.3 10^{-1}

Fig. 6.31 Affine invariant global recognition: all pairs of curves showing matches with NFA < 1. Right column: the MSERs extracted from the curves that match with the lowest NFA are displayed, superimposed

6.3.3 Global Matches of Non-Locally Encoded LLDs

The main drawback of global shape matching is its sensitivity to occlusion, whereas local matching is especially designed to deal with it. Nevertheless, the semi-local encoding presented in Chap. 4 is unable to encode curves which are convex or quasi-convex (curves for which the length after normalization is not large enough to be encoded). While in general (as will be shown with some experiments) these quasi-convex boundaries are not very discriminatory because they are not rich in details,

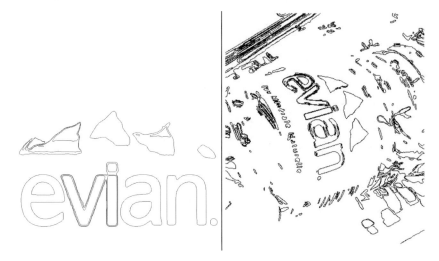

Fig. 6.32 Evian: maximal meaningful boundaries. Left: query. Right: scene

Fig. 6.33 Evian: global similarity invariant matching. All 1-meaningful matches with character 'n' from the query image. The query 'n' is represented with 10 MSERs, that match with 36 MSERs from the scene image. The lowest NFA is 10^{-11}. False detections show NFAs between 0.7 and 1

some of them may provide useful information that should not be missed. Indeed individually speaking each match may not be very meaningful, yet the conjunction of several of them can become very significant. Thus, the global and semi-local methods must work together. Of course the non-locally encoded LLDs are globally encoded and therefore globally compared.

Fig. 6.34 Evian: affine invariant global matching. Meaningful matches with character 'n' from the query image represented with 10 MSERs. Left: 1-meaningful matches, 35 matches. False matches show an NFA between 10^{-3} and 1, but some of them are not really false but rather casual, since they correspond to other characters 'n' and 'u' which are present in the scene. Good matches show NFA ranging from 10^{-15} to 10^{-8}. Right image: the 23 meaningful matches showing NFAs below 10^{-2}

6.3.3.1 First Example: a Book Cover

Figure 6.37 shows two different views of a book cover and its corresponding maximal meaningful boundaries. The query image (on top) consists of a partial view, taken from a different viewpoint. The two images are related by a strong perspective deformation. Perspective transformations can be locally approximated by affine transformations. Indeed many boundaries in images are quite local. It is therefore sound to try to find correspondences between the considered pair of images using the semi-local or the global affine invariant recognition methods.

Figure 6.38 shows the 1-meaningful matches between LLDs detected by the semi-local affine recognition method. Among the 16 matches a single false match with NFA equal to 0.6 can be seen on the right part of the scene image. The lowest NFA was 10^{-10}.

The next stage of the matching procedure is finding matches between MSERs extracted from those maximal meaningful boundaries that were not described by any semi-local LLD. All not semi-locally encoded LLDs are shown in Fig. 6.39(a). These two sets of curves are used as input for the global affine invariant recognition method. Figure 6.39(b) shows the detected 1-meaningful matches between MSERs. Good matches reach NFAs as low as 10^{-10}. Some false matches are detected, but they are only false because they do not correspond to the same objects. These false matches correspond to MSERs that look actually alike. Such false correspondences often occur. Convex or quasi-convex shapes are not very discriminatory. Higher

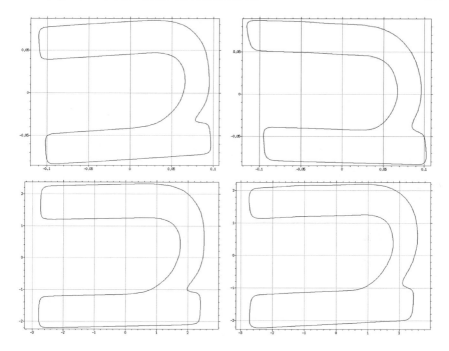

Fig. 6.35 Evian experiment. Top row: matches for the 'n' showing the lowest NFA for the global similarity invariant recognition. Bottom row: best match for the global affine invariant recognition method. In each of the rows, the curve on the left is the normalized MSER extracted from the query 'n', and the one on the right is the corresponding normalized MSER extracted from the scene image. The NFA for the similarity method was 10^{-11}, and for the affine method was 10^{-15}. In spite of the projection on the bottle, the normalized MSERs are very alike

level information (such as spatial coherence between matches) is needed in order to assess their semantic validity.

Notice that if we combine the matches that were obtained from both the semi-local and affine invariant methods, almost all shapes in common are detected. Compare now the combination of these matches with the matches detected when using the global method over all the LLDs (Fig. 6.40). Using first the semi-local method and then the global method over the non semi-locally encoded LLDs produces fewer false matches than using the global method over the original sets of LLDs. Even when not dealing with occlusions considering semi-local descriptions for complicated boundaries is more efficient than describing them globally.

6.3.3.2 Two Frames of a Sequence

Figure 6.41 shows the semi-locally matched LLDs between two frames of a movie sequence using the semi-local similarity invariant method. The non semi-locally encoded LLDs are displayed in Fig. 6.42. The majority of the non semi-locally

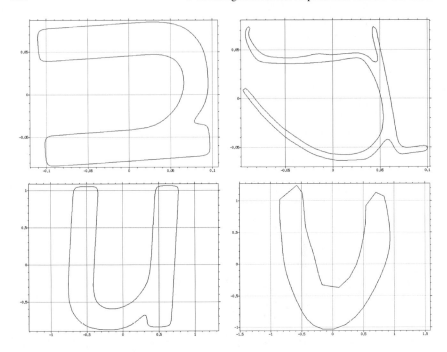

Fig. 6.36 Evian experiment. Top row: the false match for the 'n' that shows the lowest NFA, for the global similarity invariant recognition method. Bottom row: best match for the affine invariant recognition method. In each of the rows, the curve on the left is the normalized MSER extracted from the query 'n', and the one on the right is the corresponding normalized MSER extracted from the scene image. The NFA for the similarity invariant match was 0.7, and for the affine method $4.0\,10^{-3}$ (seen in the handwriting on the top of the right image from Fig. 6.34)

encoded LLDs are oval shaped and not discriminatory enough to decide if a match is semantically correct. Nevertheless, while pairing two of them may not provide much information, looking for spatial coherence between all pairs of matches can lead to high confidence detections.

Figure 6.43 shows some MSER matches (those for which the NFA is below 10^{-2}). Almost all represented MSERs seem to be discriminatory enough. No oval shaped boundary is present. Such shapes are rarely discriminatory. This fact is consistent with the detection methodology: good matches between discriminatory shape elements show the lowest NFAs.

Fig. 6.37 Book cover. Top row: query image, and its corresponding 208 maximal meaningful boundaries. Bottom row: scene image, and its 1,185 maximal meaningful boundaries

Fig. 6.38 Book cover: the 16 semi-local affine invariant matches. The NFA of the best match is 10^{-10}. The scene LLD of the only false match that was detected (NFA = 0.6) can be seen on the right part of the scene image

Fig. 6.39 Book cover. Top row: all not locally encoded LLDs. Too small or too convex level lines are not encoded. Bottom row: the 160 matches between MSERs, using the global affine invariant recognition method. The search is only performed on the LLDs which had not been locally encoded. The NFA's of some matches reach values as low as 10^{-10}. Since spatial coherence between matches is not taken into account, false matches (from a semantic viewpoint) are unavoidable (these matches correspond to MSERs that actually are alike)

Fig. 6.40 Book cover. The 857 global LLDs (MSERs) detected as 1-meaningful matches among all LLDs. The lowest NFA reaches 10^{-14}. The majority of the false matches are unavoidable since the MSERs are very alike

Fig. 6.41 Movie frames. The 75 semi-local similarity invariant 1-meaningful matches. The lowest NFA is about $2.0\,10^{-16}$

Fig. 6.42 Movie frames. Non semi-locally encoded maximal meaningful boundaries. There are 356 lines in the query image (left) and 373 in the scene image (right)

Fig. 6.43 Movie frame. The 120 global 10^{-2}-meaningful matches among the non semi-locally encoded LLDs. The lowest NFA is about $5.0\,10^{-13}$

Part IV
Grouping Shape Elements

Chapter 7
Hierarchical Clustering and Validity Assessment

Abstract The unsupervised classification of points into groups is commonly referred to as *clustering* or *grouping*. Clustering aims at discovering structure in a point data set by dividing it into its natural groups. There are three classical problems related to the construction of the right clusters. The first is evaluating the *validity* of a cluster candidate. In other words, is a group of points really a cluster, i.e. a group with a large enough density? The second problem is that meaningful clusters can contain or be contained in other meaningful clusters. A rule is needed to define locally optimal clusters by inclusion. This rule, however, is not enough to interpret correctly the data. The third problem is defining a correct merging rule between meaningful clusters, and thus being able to decide whether they should stay separate or unit. A unified *a contrario* method will be proposed for these problems. In continuation, some complexity issues and heuristics to find sound candidate clusters will be considered. In the next chapters, the clustering theory developed here will find a main application in shape recognition: the grouping of several local matches into a more global shape matching.

7.1 Clustering Analysis

The previous chapters proved that it is possible to define shape elements in images with invariance properties that agree with visual perception. This definition is accurate in the sense that two random shape elements look similar with a very small probability in the *ad hoc* background model. This is, however, only the first stage in the shape identification process. The next stage should assert whether several shape elements belong to the same shape or not. Shape elements must be grouped into what would be more properly called a shape. In this chapter, the grouping problem will be addressed in a very general setting where the problem is to group data points in a metric space. The next chapter will develop the particular application to shape element grouping. The point data set will then be specific. Each point will be a geometric transformation (similarity or affine transformation) predicted by a matching

F. Cao et al., *A Theory of Shape Identification*. Lecture Notes in Mathematics 1948.
© Springer-Verlag Berlin Heidelberg 2008

pair of shape elements. Each cluster of transformations will correspond to a globally recognized shape.

The classification of general data into groups is usually referred to as clustering. Let $E \subset \mathbb{R}^D$ and consider a data set $\mathcal{D} = \{x_1, \ldots, x_M\}$ of M points in E (some of them may be equal). The clustering problem consists in finding disjoints groups $G_1, \ldots, G_k$ with $\cup_{i=1}^{k} G_i \subset \mathcal{D}$. The inclusion can a priori be strict; the G_i may not form a partition of $\mathcal{D}$. Of course, in order to give a quantitative relevance to each group, E is equipped with a dissimilarity function $d : E \times E \to \mathbb{R}_+$. The groups are then constructed so that each one contains homogeneous data (intra-cluster similarity) and the content of different groups is fairly different (inter-cluster dissimilarity). Many techniques for achieving this goal have been proposed. The gentle reader is referred to the Keynotes A.1 for a short review and useful references. The class of method to be used depends on the problem and the type of data to be processed.

Still, there are three main general problems associated with cluster detection (see Fig. 7.1):

1. Cluster validity: How to assess the relevance of a group of data points? A validity or meaningfulness measure should be defined.
2. Optimization: How to find relatively exact borders for each group?
3. Merging rule: When two valid clusters are included in another one, is it better to merge them or to keep them separate?

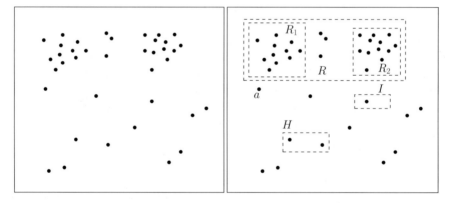

Fig. 7.1 This figure illustrates three aspects of the grouping problem. The figure presents a set of data points in the plane and some test regions where an exceptional density may be observed or not. Intuitively, regions H and I do not contain clusters. So the first question is to rule out such non meaningful clusters. A second question is the choice of sound candidate regions: for instance, should R_1 be enlarged to include the point a? Another problem is to find the best description of the observed clusters. The region R is a possible good candidate, but it also contains the points of regions R_1 and R_2 which also are sound candidates. Thus, the question arises of whether R should be chosen as cluster region, rather than the pair (R_1, R_2)

This chapter describes an a contrario decision method to answer these three questions.

7.2 *A Contrario* Cluster Validity

7.2.1 The Background Model

In everything that follows $E \subset \mathbb{R}^D$ is endowed with a probability measure π (which will also be called *background law*). By definition, $\pi(R)$ is the probability that a random point belongs to R. We do not mention measurability issues here. They are straightforward enough in this context.

The definition of π is problem specific. In general it is given *a priori* or can be empirically estimated over the data (see next chapter).

Definition 12. A *point background process* is a finite point process $(X_i)_{i=1, \ldots M}$ in E made of M mutually independent variables, identically distributed with law π.

A standard way to construct such a point process from (E, π) is to consider the product probability space $(E^M, \Pr = \pi^M)$ and the random variables X_i defined by $X_i(x) = x_i$ for any $x = (x_i)_{i=1, \ldots, M} \in E^M$.

Let us now consider an observed data set of M points $\{x_1, \ldots, x_M\}$ in E. Exactly as in Chap. 5, a subset of the data set will form a meaningful group if it could not occur by chance. In other words, it could not be explained by the background model. Therefore, the cornerstone of the *a contrario* method is to contradict the following assumption:

(A) *The observed M-tuple* $(x_i)_{i \in \{1 \ldots M\}}$ *is a realization of the background process.*

Let us give an example to illustrate this idea. Figure 7.2 represents two 2D projections of a 4-dimensional set of points. These points correspond to similarities applying a shape element in an image to the matched shape element in another image by the method described in Chap. 8. The high density of a region of the space reveals that the points therein correspond to the same shape. The probability that such a concentrated cluster is a realization of the background process is very low.

It is assumed that an agglomeration algorithm is given. This is defined as a function

$$\mathcal{A}: \quad E^M \to (\mathcal{P}(E))^P$$
$$(x_1, \ldots, x_M) \mapsto \mathcal{A}(x_1, \ldots, x_M) = (G_1, \ldots, G_P) \tag{7.1}$$

which to any M-tuple of data points associates a P-tuple of sets, $G_1, \ldots, G_P$, such that each G_k is a part of $\{x_1, \ldots, x_M\}$. The algorithm $\mathcal{A}$ is designed from any clustering algorithm and proposes a set of group candidates from a set of data points. The number of candidates P only depends on the number of data points M and not on the particular values of $x_1, \ldots, x_M$. A particular choice for $\mathcal{A}$ will be given in Sect. 7.4, but all the theory hereafter does not depend on this choice. Some of the candidate groups may actually be empty, meaning that P is an upper bound of the number of possible groups.

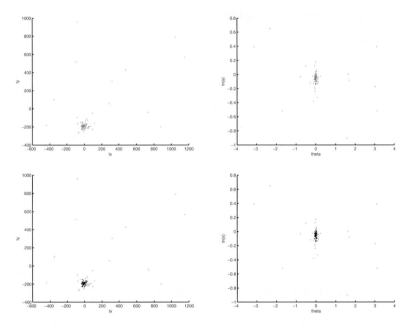

Fig. 7.2 Top row: two projections of a 4-dimensional data point set corresponding to a problem studied in Chap. 8, Fig. 9.15. Each dot represents a similarity associated with a meaningful match between shape elements. A group of dots corresponds to a coherent set of similarities indicating that the matched shapes belong to the same global shape. Thus, optimally assembling shapes reduces to the search of optimal point clusters in dimension 4. The question of finding the right groups is crucial. Errors can lead to adding spurious elements or to removing correct elements from a shape. The two top plots are the raw data to be clustered. Bottom row: these plots depict (in black dots) the only group detected by the method presented in this chapter

7.2.2 Meaningful Groups

Consider a small region $R \subset E$ containing the origin, typically a hyperrectangle centered at the origin. Assume that k points among $(x_1, \ldots, x_M)$ belong to a region of the type $x_j + R$, for some j, $1 \leqslant j \leqslant M$. If k is large enough and R small enough one will observe a cluster of points in R which can hardly have been generated by the background model. This group of points will then be detected *a contrario* in $x_j + R$. Clusters can be grouped around any of the x_j and can have any shape. A generic shape for the tested regions must, however, be fixed *a priori*. The region R will have to belong to a finite family $\mathcal{R}$ of regions, which will be detailed later. For the time being simply assume that $\mathcal{R}$ has finite cardinality $\#\mathcal{R}$ and that $0 \in R$ for all $R \in \mathcal{R}$.

In the following, for $k \leqslant M \in \mathbb{N}$ and $0 \leqslant p \leqslant 1$, let us denote the tail of the binomial law by

$$\mathcal{B}(M, k, p) = \sum_{j \geqslant k} \binom{M}{j} p^j (1-p)^{M-j}.$$

Given a background process $X_1, \ldots, X_M$ and a region R of E with probability $\pi(R)$, $\mathcal{B}(M, k, \pi(R))$ can be interpreted as the probability that *at least k out of the M points of the process belong to R*. A thorough study of the binomial tail and its use in the detection of geometric structures is presented in [50] and the textbook [54]. In this latter book, a chapter is dedicated to clustering, but with a method which is only valid in dimension 2.

Definition 13. Let $G \subset \{x_1, \ldots x_M\}$ a subset of k points out of the M data points. We call number of false alarms of G,

$$\mathrm{NFA}_g(G) \equiv \#\mathcal{R} \cdot M \cdot P \cdot \min_{\substack{x_j \in G, R \in \mathcal{R} \\ G \subset x_j + R}} \mathcal{B}(M-1, k-1, \pi(x_j + R)). \qquad (7.2)$$

We say that G is an ε-meaningful group if $\mathrm{NFA}_g(G) < \varepsilon$.

As a sanity check of the above definition, the aim is to prove that the expected number of ε-meaningful regions is less than ε, when the data set $x_1, \ldots, x_M$ is a realization of the background process and the group candidates result from the agglomeration algorithm $\mathcal{A}$.

Careful notation is needed. Let us fix $1 \leqslant j \leqslant M$ and $R \in \mathcal{R}$. We note:

- $X = (X_1, \ldots, X_M)$, the background process;
- $x = (x_1, \ldots x_m)$ a set of M points in E;
- $X^j = (X_1, \ldots, X_M)$ with X_j omitted in the list;
- $x^j = (x_1, \ldots, x_M)$ with x_j omitted in the list;
- $d\pi^j(x^j) = d\pi(x_1) \ldots d\pi(x_M)$ with $d\pi(x_j)$ omitted in the product;
- Pr^j the marginal of Pr with respect to X^j;
- $K(X^j, X_j, R)$ number of points in the list X^j belonging to $X_j + R$.

Lemma 5. *Let us fix $x_j \in E$. Consider a random process $X_1, \ldots, X_M$. Then*

$$\mathrm{Pr}^j \left(\mathcal{B}(M-1, K(X^j, x_j, R), \pi(x_j + R)) < \frac{\varepsilon}{\#\mathcal{R} \cdot M} \right) \leqslant \frac{\varepsilon}{\#\mathcal{R} \cdot M}.$$

Proof. The repartition function of the random variable $K(X^j, x_j, R)$ is $k \mapsto \mathcal{B}(M-1, k, \pi(x_j + R))$. The result follows from Lem. 2, p. 21. $\square$

Proposition 8. *Let $X_1, \ldots X_M$ be a background process. Consider the P random groups $\mathcal{A}(X_1, \ldots, X_M) = (\Gamma_1, \ldots, \Gamma_P)$. Then the expected number of the ε-meaningful groups among $\Gamma_1, \ldots, \Gamma_P$ is less than ε.*

Proof. Note

- For $1 \leqslant i \leqslant P$, the Bernoulli variable

$$Y_i = \begin{cases} 1 \text{ if } \Gamma_i \text{ is } \varepsilon\text{-meaningful,} \\ 0 \text{ otherwise.} \end{cases}$$

- $S = \sum_i Y_i$ the number of ε-meaningful groups.

Also denote by K_i the (random) cardinality of Γ_i and $\epsilon = \frac{\varepsilon}{MP\#\mathcal{R}}$.

$$\Pr(Y_i = 1) = \Pr\left(\min_{\substack{X_j \in \Gamma_i, R \in \mathcal{R} \\ \Gamma_i \subset X_j + R}} \mathcal{B}(M - 1, K_i - 1, \pi(X_j + R)) < \epsilon \right) \quad (7.3)$$

$$= \Pr(\exists j, R \text{ s.t. } X_j \in \Gamma_i, \Gamma_i \subset X_j + R, \quad\quad (7.4)$$
$$\mathcal{B}(M - 1, K_i - 1, \pi(X_j + R)) < \epsilon)$$

$$\leqslant \Pr(\exists j, R \text{ s.t. } \mathcal{B}(M - 1, K(X^j, X_j, R), \pi(X_j + R)) < \epsilon) \, (7.5)$$

$$\leqslant \sum_{\substack{1 \leqslant j \leqslant M \\ R \in \mathcal{R}}} \Pr(\mathcal{B}(M - 1, K(X^j, X_j, R), \pi(X_j + R)) < \epsilon). \quad (7.6)$$

The first inequality results from $\Gamma_i \subset X_j + R \Rightarrow K_i - 1 \leqslant K(X^j, X_j, R)$ and the monotonicity of the map $k \mapsto \mathcal{B}(M - 1, k, p)$. Now, Lem. 5 cannot be directly applied. Indeed, the considered region is centered at a random point X_j and thus has a random probability. However, by Fubini Theorem

$$\Pr(\mathcal{B}(M - 1, K(X^j, X_j, R), \pi(X_j + R)) < \epsilon)$$
$$= \int d\pi(x_j) \Pr{}^j (\mathcal{B}(M - 1, K(X^j, x_j, R), \pi(x_j + R)) < \epsilon),$$
$$\leqslant \int d\pi(x_j)\epsilon \quad \text{by Lem. 5,}$$
$$= \epsilon.$$

Thus

$$\Pr(Y_i = 1) \leqslant M\#\mathcal{R}\epsilon = \frac{\varepsilon}{P}.$$

Finally,

$$\mathbb{E}(S) = \sum_{i=1}^{P} \mathbb{E}(Y_i) \leqslant \sum_{i=1}^{P} \frac{\varepsilon}{P} = \varepsilon. \qquad \square$$

Remark 5. As in Chap. 2 and 5, the key point is that the *expectation* of the number S of meaningful regions is easily controlled. The probability law of S would instead be extremely difficult to compute because of the interaction between regions.

To summarize: The number of false alarms is a measure of how likely it is that a group G containing at least k of the data points, was generated by chance, as a realization of the background process. The lower $\mathrm{NFA}_g(G)$, the less likely the observed cluster in the background process. By Prop. 8, the only parameter controlling the detection is ε. This provides a handy way to control false detections. If, on average, one is ready to tolerate one non relevant region among all regions, then ε can be simply set to 1.

The following proposition shows that the influence of the parameter $\#\mathcal{R}$ and of the decision parameter ε on the detection results are very weak.

Proposition 9 ([50]). *Let* $0 < p < 1$ *and*

$$k^* = \min\{k : MP\#\mathcal{R} \cdot \mathcal{B}(M - 1, k, p) \leqslant \varepsilon\}.$$

Then

$$\alpha\sqrt{2p(1 - p)} \leqslant k^* - p(M - 1) \leqslant \frac{\alpha}{\sqrt{2}}, \qquad (7.7)$$

where $\alpha = \sqrt{(M - 1)\ln(MP\#\mathcal{R}/\varepsilon)}.$

Notice that k^* is the minimal number of points in a ε-meaningful group and thus depends on the size of the regions containing the group. By the preceding result, this decision threshold only has a logarithmic dependence upon P, $\#\mathcal{R}$ and ε.

Figure 7.3 shows an example of clustering. The data consists of 950 points uniformly distributed in the unit square, and 50 points manually added around the positions $(0.4, 0.4)$ and $(0.7, 0.7)$. The figure shows the result of a numerical method involving the above NFA. Both visible clusters are found with NFA_g respectively equal to 10^{-8} and 10^{-7}. Such low numbers can hardly be the result of chance. How to obtain *exactly* these two clusters and no other larger or smaller ones which would also be meaningful? This will be discussed in the next two sections.

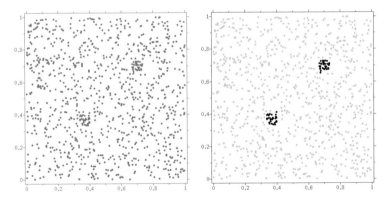

Fig. 7.3 Clustering of twice 25 points around $(0.4, 0.4)$ and $(0.7, 0.7)$ surrounded by 950 i.i.d. points, uniformly distributed in the unit square. The regions of $\mathcal{R}$ are rectangles as described in Sect. 7.4.1. In this example $\#\mathcal{R} = 2500$ (50 different sizes in each direction). Exactly two maximal meaningful clusters are detected. The NFA_g of the lower left one is 10^{-8} while the upper-right one has a NFA_g equal to 10^{-7}

7.3 Optimal Merging Criteria

7.3.1 Local Merging Criterion

While each meaningful group is relevant by itself, the whole set of meaningful regions exhibits in general a high redundancy. Indeed a very meaningful group G usually remains meaningful when it is slightly enlarged or shrunk into a group G'. If, e.g. $G \subset G'$, this question is easily answered by a comparing $\text{NFA}_g(G)$ and $\text{NFA}_g(G')$. The group with the smallest number of false alarms must of course be preferred. Another more subtle question arises when three or more groups interact. Let G_1 and G_2 be two tested disjoint groups and G another tested group containing $G_1 \cup G_2$. We then face two conflicting interpretations of the data: two clusters or just one? The merged group G is not necessarily a better data representation than the two separate clusters G_1 and G_2. One possibility is that G is less meaningful than each one of the merging groups. In such a case, G_1 and G_2 should be kept rather than G. The situation is less obvious when G is more meaningful than both G_1 and G_2. In this case, keeping G_1 and G_2 separated may still be opportune. So a quantitative merging criterion is required. We shall first define a *number of false alarms for a pair of groups*. This new value will be compared to the NFA_g of the merged group. Let us introduce the trinomial coefficient

$$\binom{M}{i,j} = \binom{M}{i}\binom{M-i}{j}.$$

We note

$$\mathcal{M}(M, k_1, k_2, \pi_1, \pi_2) = \sum_{i=k_1}^{M} \sum_{j=k_2}^{M-i} \binom{M}{i,j} \pi_1^i \pi_2^j (1 - \pi_1 - \pi_2)^{M-i-j}. \qquad (7.8)$$

This number can be interpreted as follows. Let R_1 and R_2 be two disjoint regions of E and $\pi_1 = \pi(R_1)$, $\pi_2 = \pi(R_2)$ their probabilities. Then $\mathcal{M}(M, k_1, k_2, \pi_1, \pi_2)$ is the probability that at least k_1 among the M, and then at least k_2 points among the remaining ones belong to R_1 and R_2 respectively. Thus, this probability measures how exceptional a pair of concentrated clusters can be in the background model.

As in the case of single regions, it is assumed that a set of P *pairs* of group candidates are obtained by an operator $\mathcal{A}_2$. That is to say

$$\mathcal{A}_2: \qquad E^M \to (\mathcal{P}(E) \times \mathcal{P}(E))^P \qquad (7.9)$$
$$(x_1, \ldots, x_M) \mapsto \mathcal{A}_2(x_1, \ldots, x_M) = ((G_1^1, G_1^2), \ldots, (G_P^1, G_P^2)),$$

where it is assumed that $G_i^k \subset \{x_1, \ldots x_M\}$, for $k = 1, 2$ and $1 \leqslant i \leqslant P$.

Definition 14. Consider two group candidates (G^1, G^2) of data points. Let $(z_1, z_2) \in G^1 \times G^2$ be two data points, and R_1 and R_2 in $\mathcal{R}$. Let us denote by

- k_1 (resp. k_2) the cardinality of $G^1 \setminus (z_2 + R_2)$ (resp. $G^2 \setminus (z_1 + R_1)$), i.e. the number of points of G^1 (resp. G^2) that are not in $z_2 + R_2$ (resp. $z_1 + R_1$).
- $\pi_1 = \pi((z_1 + R_1) \setminus (z_2 + R_2))$ and $\pi_2 = \pi((z_2 + R_2) \setminus (z_1 + R_1))$.

Let us define the number of false alarms of the pair (G^1, G^2) by

$$\text{NFA}_{gg}(G^1, G^2) = M^3 \cdot P \cdot (\#\mathcal{R})^2 \min_{\substack{(z_1, z_2) \in G^1 \times G^2, \\ R_1, R_2 \in \mathcal{R} \\ G^1 \subset z_1 + R_1, \\ G^2 \subset z_2 + R_2}} \mathcal{M}(M-2, k_1-1, k_2-1, \pi_1, \pi_2).$$

(7.10)

We say that a pair of groups (G^1, G^2) is ε-meaningful if $\text{NFA}_{gg}(G^1, G^2) < \varepsilon$.

Let us sum up how to compute this quantity. Choose a region centered at one point of G_1 (resp. G_2) and containing G_1 (resp. G_2). Those two regions may intersect, so remove their intersection and the points it may contain. Then, k_1 and k_2 points are left in each group and the trinomial tail can be computed. As above, this quantity measures how likely it is that G_1 and G_2 contain *simultaneously* at least k_1 and k_2 points. Removing the intersection is a mere technicality and the probability of this event is the tail of the trinomial law.

As usual, the aim is to prove that the expected number of ε-meaningful pairs of regions is less than ε. As in the study of ε-meaningful groups, some care must be taken of notations and abbreviations. Let $1 \leqslant i \neq j \leqslant M$. Two tested regions $x_i + R_i$ and $x_j + R_j$ may intersect and we have to deal with this possibility. (The indices i and j in R_i and R_j only aim at reminding us that these are regions centered at x_i and x_j although this notation is actually a bit incorrect). We note

- $X = (X_1, \ldots, X_M)$, the background process,
- $x = (x_1, \ldots x_M)$ a set of M dots in E,
- $X^{ij} = (X_1, \ldots, X_M)$ with X_i, X_j omitted in the list,
- $x^{ij} = (x_1, \ldots, x_M)$ with x_i, x_j omitted in the list,
- $X_{ij} = (X_1, \ldots, X_M)$ with X_i and X_j replaced by x_i and x_j,
- $d\pi^{ij}(x^{ij}) = d\pi(x_1) \ldots d\pi(x_M)$ with $d\pi(x_i)$ and $d\pi(x_j)$ omitted in the product,
- Pr^{ij} the marginal of Pr with respect to x^{ij},
- $K(X, i, j, R_i, R_j) = $ the number of points among X^{ij} that are in $X_i + R_i$ but not in $X_j + R_j$, i.e. belonging to $(X_i + R_i) \setminus (X_j + R_j)$,
- $K_i = K(X, i, j, R_i, R_j)$, $K_j = K(X, j, i, R_j, R_i)$,
- $\tilde{K}_i = K(X_{ij}, i, j, R_i, R_j)$, $\tilde{K}_j = K(X_{ij}, j, i, R_j, R_i)$,
- $k_i = K(x, i, j, R_i, R_j)$, $k_j = K(x, j, i, R_j, R_i)$,
- $\pi_i = \pi((x_i + R_i) \setminus (x_j + R_j))$, $\pi_j = \pi((x_j + R_j) \setminus (x_i + R_i))$,
- $\Pi_i = \pi((X_i + R_i) \setminus (X_j + R_j))$, $\Pi_j = \pi((X_j + R_j) \setminus (X_i + R_i))$,
- $\epsilon = \frac{\varepsilon}{M^3 P (\#\mathcal{R})^2}$.

Lemma 6. *For every* $x_i, x_j \in E$,

$$\text{Pr}^{ij} \left[\mathcal{M}(M - 2, \tilde{K}_i, \tilde{K}_j, \pi_i, \pi_j) < \epsilon \right] < (M - 1)\epsilon.$$

Proof. The proof extends the arguments used for Lem. 2, p. 21 to the case of two variables. Notice that this proof is true for discrete variables since it uses the fact that $\tilde{K}_j$ and $\tilde{K}_i$ can only take $M-1$ different values. Indeed,

$$
\begin{aligned}
&\mathrm{Pr}^{\,ij}\left[\mathcal{M}(M-2,\tilde{K}_i,\tilde{K}_j,\pi_i,\pi_j)<\epsilon\right]\\
&=\sum_{(k_i,k_j)\mid \mathcal{M}(M-2,k_i,k_j,\pi_i,\pi_j)<\epsilon}\mathrm{Pr}^{\,ij}(\tilde{K}_i=k_i,\tilde{K}_j=k_j)\\
&=\sum_{(k_i,k_j)\mid \mathcal{M}(M-2,k_i,k_j,\pi_i,\pi_j)<\epsilon}\binom{M-2}{k_i,k_j}\pi_i^{k_i}\pi_j^{k_j}(1-\pi_i-\pi_j)^{M-2-k_i-k_j}.
\end{aligned}
$$

Let

$$
k_i(\epsilon,k_j)=\inf\{0\leqslant k\leqslant M-2\mid \mathcal{M}(M-2,k,k_j,\pi_i,\pi_j)<\epsilon\},
$$

with the useful conventions $\mathcal{M}(M-2,k,k_j,\pi_i,\pi_j)=0$ and $\binom{M-2}{k,k_j}=0$ if $k\geqslant M-1-k_j$. The map $k\mapsto\mathcal{M}(M-2,k,k_j,\pi_i,\pi_j)$ being monotone,

$$
\mathcal{M}(M-2,k,k_j,\pi_i,\pi_j)<\epsilon\Leftrightarrow k\geqslant k_i(\epsilon,k_j).\tag{7.11}
$$

Summarizing and using the definition of $k_i(\epsilon,k_j)$,

$$
\begin{aligned}
&\mathrm{Pr}^{\,ij}\left[\mathcal{M}(M-2,\tilde{K}_i,\tilde{K}_j,\pi_i,\pi_j)<\epsilon\right]\\
&=\sum_{k_j=0}^{M-2}\sum_{k=k_i(\epsilon,k_j)}^{M-2}\binom{M-2}{k,k_j}\pi_i^{k}\pi_j^{k_j}(1-\pi_i-\pi_j)^{M-2-k-k_j}\\
&\leqslant\sum_{k_j=0}^{M-2}\sum_{k=k_i(\epsilon,k_j)}^{M-2}\sum_{l=k_j}^{M-2-k}\binom{M-2}{k,l}\pi_i^{k}\pi_j^{l}(1-\pi_i-\pi_j)^{M-2-k-l}\\
&=\sum_{k_j=0}^{M-2}\mathcal{M}(M-2,k_i(\epsilon,k_j),k_j,\pi_i,\pi_j)<(M-1)\epsilon.\qquad\square
\end{aligned}
$$

Proposition 10. *Consider a background process $X_1,\ldots,X_M$ and the P random pairs $\mathcal{A}_2(X_1,\ldots,X_M)=((\Gamma_1^1,\Gamma_1^2),\ldots,(\Gamma_P^1,\Gamma_P^2))$. Then the expected number of ε-meaningful pairs of regions among them is less than ε.*

Proof. Let us note for $k=1,\ldots,P$

- The Bernoulli variable

$$
Y_k=\begin{cases}1 \text{ if } (\Gamma_k^1,\Gamma_k^2) \text{ is } \varepsilon\text{-meaningful,}\\[2mm]0 \text{ otherwise.}\end{cases}
$$

- $S=\sum_{k=1}^{P}Y_k$ the number of ε-meaningful pairs of regions.

Let us fix k. Let X_i and X_j two points in the process, belonging to Γ_k^1 and Γ_k^2. Let R_i and R_j be two regions in $\mathcal{R}$, such that $\Gamma_k^1 \subset X_i + R_i$ and $\Gamma_k^2 \subset X_j + R_j$. Let also $\hat{K}_i$ be the number of points of Γ_k^1 that are not in $X_j + R_j$ and $\hat{K}_j$ the number of points of Γ_k^2 that are not in $X_i + R_i$. Notice that with the notations above, $\hat{K}_i - 1 \leqslant K_i$ and $\hat{K}_j - 1 \leqslant K_j$. Then,

$$
\begin{aligned}
\Pr(Y_k = 1) = \Pr(&\exists i, j, R_i, R_j \quad \text{s.t. } X_i \in \Gamma_k^1, X_j \in \Gamma_k^2, \\
&\Gamma_k^1 \subset X_i + R_i, \Gamma_k^2 \subset X_j + R_j, \\
&\mathcal{M}(M - 2, \hat{K}_i - 1, \hat{K}_j - 1, \Pi_i, \Pi_j) < \epsilon). \\
\leqslant \Pr(&\exists i, j, R_i, R_j \quad \text{s.t. } \mathcal{M}(M - 2, K_i, K_j, \Pi_i, \Pi_j) < \epsilon) \\
\leqslant \sum_{i,j=1}^{M} \sum_{R_i, R_j} &\Pr(\mathcal{M}(M - 2, K_i, K_j, \Pi_i, \Pi_j) < \epsilon)
\end{aligned}
$$

The first inequality results from $\hat{K}_i - 1 \leqslant K_i$ and $\hat{K}_j - 1 \leqslant K_j$ and the monotonicity of the map $(k, l) \mapsto \mathcal{M}(M - 2, k, l, p, q)$ with respect to each of its variables. By Fubini's theorem,

$$
\begin{aligned}
\Pr(\mathcal{M}(M - 2, K_i, K_j, \Pi_i, \Pi_j) < \epsilon) \\
= \int_{E^2} d\pi(x_i) d\pi(x_j) \int_{E^{M-2}} \mathbb{1}_{\{\mathcal{M}(M-2, k_i, k_j, \pi_i, \pi_j) < \epsilon\}} d\pi^{ij}(x^{ij}) \\
= \int_{E^2} d\pi(x_i) d\pi(x_j) \Pr{}^{ij}(\mathcal{M}(M - 2, \tilde{K}_i, \tilde{K}_j, \pi_i, \pi_j) < \epsilon) \\
< (M - 1)\epsilon,
\end{aligned}
$$

where Lem. 6 has been used in the last inequality. Finally,

$$
\begin{aligned}
\mathbb{E}(S) = \sum_{k=1}^{P} \mathbb{E}(Y_k) \\
< \sum_{k=1}^{P} (M - 1)M^2(\#\mathcal{R})^2 \epsilon \\
\leqslant \varepsilon. \qquad \square
\end{aligned}
$$

Definition 15 (Merging condition). Let G_1 and G_2 be two groups and G containing $G_1 \cup G_2$. We say that G is indivisible relatively to G_1 and G_2 if

$$
\text{NFA}_g(G) \leqslant \text{NFA}_{gg}(G_1, G_2). \tag{7.12}
$$

Equation (7.12) represents a crucial test for the coherence of a cluster. If it is not fulfilled, G will not be considered a valid region as it can be divided into a more meaningful pair of cluster regions. The next lemma will prove useful to give a very coarse, but qualitatively handy characterization for the relationship between the NFA$_g$s of a cluster, and two disjoint clusters contained in it.

Lemma 7. *For every k_1 and k_2 in $\{0, \ldots, M\}$, such that $k_1 + k_2 \leqslant M$ and for every π_1 and π_2 in $[0, 1]$ such that $\pi_1 + \pi_2 \leqslant 1$,*

$$\mathcal{M}(M, k_1, k_2, \pi_1, \pi_2) \leqslant \mathcal{B}(M, k_1, \pi_1) \cdot \mathcal{B}(M, k_2, \pi_2). \tag{7.13}$$

A proof of the lemma is given in Appendix A.3. We are actually interested in its consequence.

Proposition 11. *Let G be indivisible with respect to G_1 and G_2. Let also assume that the regions related to G_1 and G_2 are disjoint. Then*

$$\mathrm{NFA}_g(G) < \frac{M}{P} \cdot \mathrm{NFA}_g(G_1) \cdot \mathrm{NFA}_g(G_2)$$

Proof. Let us denote by π_1 and π_2 the probability of the regions attaining the NFA_g of G_1 and G_2. These regions are assumed to be disjoint. By an obvious monotonicity argument those regions also attain the minimum of the trinomial law. From (7.10) and Def. 15,

$$\mathrm{NFA}_{gg}(G_1, G_2) = M^3 P (\#\mathcal{R})^2 \mathcal{M}(M - 2, k_1 - 1, k_2 - 1, \pi_1, \pi_2)$$

and

$$\mathrm{NFA}_g(G_i) = M P (\#\mathcal{R}) \mathcal{B}(M - 1, k_i - 1, \pi_i), \ i = 1, 2.$$

By Lem. 7, it follows that

$$\mathrm{NFA}_{gg}(G_1, G_2) \leqslant M^3 P \cdot (\#\mathcal{R})^2 \cdot \mathcal{B}(M - 2, k_1 - 1, \pi_1) \mathcal{B}(M - 2, k_2 - 1, \pi_2).$$

Since $\mathcal{B}(M - 2, k - 1, p) \leqslant \mathcal{B}(M - 1, k - 1, p)$ for all M, k and p, the result follows. $\square$

Proposition 11 yields a very simple necessary condition for cluster indivisibility. However the bound is not tight enough to be used as a merging condition: in practice many clusters G satisfy this condition without being indivisible (for an example, see Fig. 7.5). Nevertheless, the condition may still be useful from a qualitative point of view. An example of its use will be shown in the next chapter.

7.4 Computational Issues

7.4.1 Choosing Test Regions

What is the right set of test regions $\mathcal{R}$? This question is obviously application driven. To fix ideas, let us just indicate a sound and generic choice. For some reasonably fixed $a > 0$, $r > 1$ and $n \in \mathbb{N}$, consider all hyperrectangles whose edge lengths belong to the set $\{a, ar, ar^2, \ldots, ar^n\}$. This allows one to consider a tractable number of test regions with very different sizes and shapes. The choice of the hyperrectangles is particularly opportune when the probability distribution π defined on a hyperrectangle E of $\mathbb{R}^D$ is a tensor product of one-dimensional densities $\pi_1, \ldots, \pi_D$. We address the question with more details in the next chapter.

Definition 13 permits us to compute the NFA_g of any group of points. This computation involves a region centered at a data point. Since the number of scales is n in each dimension, there are Mn^D regions centered at a data point. In the next chapter $D = 4$ or 6. From the numerical feasibility viewpoint, Mn^D becomes too large when n grows. Hence, detection cannot be performed by scanning all the regions centered at a point, counting the number of points it contains and computing the tail of the binomial law. The agglomeration procedure $\mathcal{A}$ involved in the definition of the meaningful groups precisely aims at reducing the number of test groups. In the experiments this agglomeration algorithm is a classical hierarchical clustering algorithm. It provides a binary tree structure from a data set $(x_1, \ldots, x_M)$. This tree, sometimes called *dendrogram*, contains exactly $2M - 1$ nodes, M of which are singletons. The pairs of $\mathcal{A}_2$ are simply obtained as the children of a node of this binary tree. There are at most $P = M - 1$ of them.

More generally speaking, hierarchical clustering methods provide a family of nested partitions of the point data set that can always be embedded in a tree structure (that may not be binary).

Sect. A.1 describes some of the main aggregation techniques for building such trees. Many of them perform a recursive binary merging procedure. Thus, they directly yield binary trees. In such methods the initial set of nodes is the set of data singletons, $\{x_1\}, \ldots, \{x_N\}$. At each stage of the construction, the two closest nodes are united to form a parent node. The inter-cluster distance must be chosen *ad hoc*. In the case of sparse data, one can take the minimal distance $d(x_i, x_j)$ where x_i belongs to the first cluster and x_j to the second one. The nodes of the tree are all merged parts at all levels and the children of a node are the parts it was merged from. The result much depends on the choice of the dissimilarity function between clusters, for which there is no universal choice.

Such a construction introduces some needed arbitrariness. Indeed, the set of all possible partitions of a data point set is huge. A tree structure reduces the exploration to searching an optimal subtree of the initial tree structure. This reduction makes sense if the set of nodes of the initial tree structure contains roughly all groups of interest. Choosing the right metric on the data point set and the right inter-cluster distance is therefore crucial.

Given a dendrogram of the data point set, the following algorithm explores all regions centered at data points and containing a dendrogram node.

Grouping Algorithm

For each node G (candidate group) with cardinality k in the clustering tree or dendrogram:

1. Set $\text{NFA}_g(G) \leftarrow +\infty$.
2. For each $x \in G$,

   a. Find the smallest region $x + R$ centered at x containing the other data points in G.

b. Set $\text{NFA}_g(G) \leftarrow \min(\text{NFA}_g(G), MP \cdot \#\mathcal{R} \cdot \mathcal{B}(M-1, k-1, \pi(x+R)))$.

7.4.2 Indivisibility and Maximality

We are now faced with Questions 2 and 3 mentioned at the beginning of this chapter. We can get many meaningful clusters by the preceding method. Their NFA_g is known. One can also compute the NFA_{gg} of a pair of clusters, and compare it roughly to the NFA_g of their union. The next definition proposes a way to select the right clusters by using the cluster dendrogram.

Definition 16 (Maximal ε-meaningful group). A node G is maximal ε-meaningful if and only if

1. $\text{NFA}_g(G) \leqslant \varepsilon$;
2. G is indivisible with respect to any pair of sibling descents;
3. For all indivisible descent G', $\text{NFA}_g(G') \geqslant \text{NFA}_g(G)$;
4. For all indivisible ascent G', either $\text{NFA}_g(G') > \text{NFA}_g(G)$ or there exists an indivisible descent G'' of G' such that $\text{NFA}_g(G'') < \text{NFA}_g(G')$.

Condition 4 implies that G can be abandoned for a larger group only if this group has not been beaten by one of its descents. Imposing Conditions 3 and 4 ensures that two different maximal meaningful groups are disjoint.

Two simple examples illustrate the critical importance of the merging condition. Figure 7.4 shows a configuration of 100 points distributed on $[0,1]^2$, and naturally grouped in two clusters G_1 and G_2. In the hierarchical structure, G_1 and G_2 are the children of $G = G_1 \cup G_2$. All three nodes are obviously meaningful since their NFA_g is much lower than 1. Their NFA_g also is lower than the NFA_g of the other groups in the dendrogram. Taking a uniform background law, it has been checked that for this particular configuration,

$$\text{NFA}_g(G_2) < \text{NFA}_g(G) < \text{NFA}_g(G_1).$$

Clearly G_1 represents an informative part of the data that should be kept and will be. Notice that G_2 is more meaningful than G and is contained in G. Thus, G would be eliminated if only the most meaningful groups by inclusion were kept. On the other hand, G is more meaningful than G_1, so that G_1 is not a local maximum of meaningfulness with respect to inclusion. So, without the notion of indivisibility, trouble would arise: G would eliminate G_1 and G_2 would eliminate G. The result would be the solution indicated in the middle column of Fig. 7.4. Actually G is neither indivisible nor satisfies Condition 3 above since it is less meaningful than the pair (G_1, G_2) and than G_2. Thus, the result of the grouping procedure yields, in accordance with the rule of Def. 16, the pair (G_1, G_2).

In [53], the above-mentioned maximality definition was proposed. It consists in picking the lowest NFA in all the tree branches. As has just been seen, this definition is not suitable here. By this definition, G_1 would have been the only maximal meaningful cluster of the tree.

Figure 7.5 illustrates another situation where the indivisibility check yields the intuitively right solution. In this example, the union G of two clusters G_1 and G_2 is more meaningful than each separate cluster. Without the indivisibility requirement, G would be the only maximal meaningful group. This would have been coherent had G_1 and G_2 been intricate enough. In the presented case, the indivisibility condition yields two clusters G_1 and G_2, since $\mathrm{NFA}_{gg}(G_1, G_2) < \mathrm{NFA}_g(G)$.

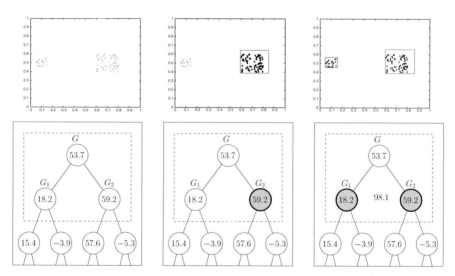

Fig. 7.4 Indivisibility prevents collateral elimination. Each subfigure shows a configuration of points and a piece of the corresponding dendrogram with the selection of maximal meaningful groups depicted in gray. The number in each node corresponds to $-\log_{10}(\mathrm{NFA}_g)$ of its associated cluster so that the cluster is meaningful when this number is large. The number placed between two nodes is the NFA_{gg} of the corresponding pair. Left: original configuration. Middle: the node selected by taking only the most meaningful group in each branch. The leftmost group G_1 is eliminated. It is, however, very meaningful since $\mathrm{NFA}_g(G_1) = 10^{-18}$. Right: by combining indivisibility and maximality criteria, both clusters G_1 and G_2 are selected

7.5 Experimental Validation: Object Grouping Based on Elementary Features

Grouping phenomena are essential in human perception since they are responsible for organizing information. In vision, grouping has been especially studied by Gestalt psychologists like Wertheimer [179]. These experiments aim at extracting the groups of objects in an image that share some elementary geometrical properties.

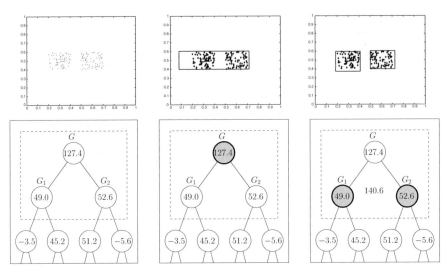

Fig. 7.5 Indivisibility prevents faulty union. Each sub-figure shows a configuration of points and a piece of the corresponding dendrogram with the selection of maximal meaningful groups depicted in gray. The number in each node corresponds to the NFA$_g$ of its associated cluster. The number between two nodes is the NFA$_{gg}$ of the corresponding pair. Left: original configuration. Middle: the node selected if one only checks maximality by inclusion and not indivisibility. The largest group G has the lowest NFA$_g$ and would be the only one kept. Note that the optimal region is not symmetric since it must be centered on a data point. Right: selected nodes obtained by combining the indivisibility and maximality criteria. Since NFA$_{gg}(G_1, G_2) = 10^{-140} < 10^{-127} = $ NFA$_g(G)$, the pair (G_1, G_2) is preferred to G

The objects boundaries are extracted as in Chap. 2. Once these objects are detected, say $O_1, ...O_M$, we can compute for each of them a list of D features (gray level, position, orientation, etc). If k objects among M have one or several features in common, one wonders if it is happening by chance or if they should be grouped. Each data point is a point in a bounded subset of $\mathbb{R}^D$ and the method described above is applied. Actually, some coordinates, such as angles, belong to the unit circle since periodicity must be taken into account. This can be done all the same.

Let us return to the experiment in Fig. 7.3. The dot process contains two groups of 25 points in addition to 950 i.i.d uniformly in the unit square. Two groups and two groups only are detected, each with very good NFA$_g$ (less than 10^{-7}).

7.5.1 Segments

In the second example, groups are perceived as a result of the collaboration between two different features. Figure 7.6 shows 71 straight segments with different orientations almost uniformly distributed in position. As expected no meaningful cluster is detected in the space of position coordinates of the barycenters. In all the experiments, the number of rectangle sizes in each direction is 50. Thus $\#\mathcal{R} = 50^D$.

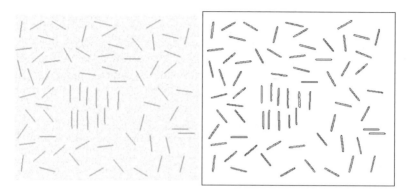

Fig. 7.6 An image of a scanned drawing of segments and its 71 maximal meaningful level lines [51]

If orientation is chosen as the only feature ($D = 1$), 6 maximal meaningful groups are detected corresponding to the most represented orientations, see Fig. 7.7. None of these clusters exhibits a very low NFA$_g$. The central group is not even detected, because the directions of the segments are slightly different. This only means that orientation is not the only perceptual grouping law used in the interpretation of this figure. All the other groups are actually not perceived because they are masked by the clutter made of all the other objects. However, they certainly have a coherent direction.

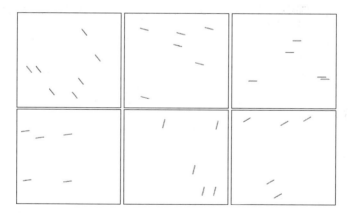

Fig. 7.7 Grouping with respect to orientation: there are 6 maximal meaningful groups. The NFA$_g$ range is between $10^{-0.4}$ and $10^{-3.8}$. Notice that the central group is missing. Indeed, the direction of the segments is not accurate and the group is not meaningful with respect to orientation. This experiment shows that orientation alone is not enough to detect some groups. Orientation is only one law of perceptual grouping among others

Let us see what happens when considering two features ($D = 2$, $\#\mathcal{R} = 2500$). In the space (x-coordinate, orientation), a single maximal meaningful cluster is found (Fig. 7.8). It corresponds to the group G of 11 central vertical segments. Its NFA_g is equal to $10^{-0.3}$, what means it is hardly meaningful with respect to these two features. In the space (y-coordinate, orientation), the combination of the maximality and the merging criterion leads to prefer the two rows of segments to the whole G. This is coherent with the visual perception since we actually see two lines of segments here. On the contrary, in the (x-coordinate, orientation) space, the merging criterion indicates that observing G is more meaningful than observing simultaneously its children in the dendrogram. This decision still conforms with observation: no particular group within G can be distinguished by the x-coordinate. The same group is obtained in the space (x-coordinate, y-coordinate, orientation), with a lower $\text{NFA}_g = 10^{-2.3}$.

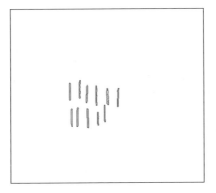

Fig. 7.8 Grouping in the space (x-coordinate,orientation). This time, the whole central group is detected and is the only maximal meaningful group. ($\text{NFA}_g = 10^{-0.3}$). If grouping is done with respect to full 2D-position and orientation, the central group is still the only detected group with $\text{NFA}_g = 10^{-2.3}$

7.5.2 DNA Image

The 80 objects in Fig. 7.9 are more challenging. More features are needed in order to represent them (diameter, elongation, orientation, etc.). It is clear that a projection on a single feature is not enough to differentiate the objects. Globally, we see three groups of objects: the DNA markers, which share the same form, size and orientation; the numbers, all on the same line, almost of the same size, and finally the elements of the ruler, also on the same line and with similar diameters. The position appears to be decisive in the perceptive formation of these groups.

In the space (diameter, y-coordinate) 6 maximal meaningful groups are detected (Fig. 7.10). Four of them correspond to the lines of DNA markers (from left to right

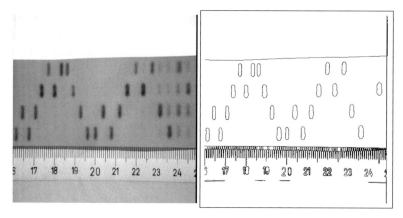

Fig. 7.9 An image of DNA and its 80 maximal meaningful level lines [51]

and top-down), $-\log_{10}(\text{NFA}_g) = 1.2, 6.1, 5.1, 4.3$. The group of numbers contains 23 objects (a group of two digits sometimes contains three objects: the two digits and a level line surrounding both of them) and $-\log_{10}(\text{NFA}_g) = 41.7$. The last group, composed of the vertical graduation of the ruler contains 31 objects and is even more meaningful: $-\log_{10}(\text{NFA}_g) = 57.3$.

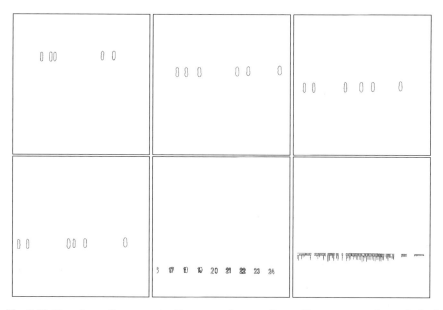

Fig. 7.10 Grouping with respect to diameter and y coordinate. Six groups are detected, 4 of which are rows of DNA markers. The last two ones correspond to the ruler. $-\log_{10}(\text{NFA}_g)$ range from 1.2 to 6.1 for the DNA. The last two groups are larger and obviously more meaningful: $-\log_{10}(\text{NFA}_g) = 41.7$ and 57.3

Assume we gave up considering the position information. Would we still see the DNA markers as a group? By taking several other features into account (see Fig. 7.11), the DNA markers form an isolated and very meaningful group: the combined features (orientation, diameter, elongation, convexity coefficient) reveal the DNA markers as a very good maximal meaningful cluster ($NFA_g = 10^{-10}$). There are two other interesting groups that are actually not detected, but whose NFA_g is also close to 1: the 1's and the 2's on the ruler.

Let us detail how π, the law of the background model, was estimated from the data themselves. The marginal distribution of each characteristic is approximated by the empirical histogram. All the characteristics are assumed to be independent. Note that the obtained distribution is not uniform at all. Why having this construction instead of a uniform law? First, there would be the assumption that the range of the data is known and it is not. Moreover, the distribution of each characteristic has no reason to be uniform. Hence, contradicting a uniform background model would produce detections caused by the discrepancy of the real distribution with respect to the uniform one. Thus we must define a background law which is as realistic as possible for each single observation. Let us now take the opposite view and ask why we did not directly take the joint empirical law. Because of sparse samples it is simply not possible to estimate this distribution. We would need at least one million data points. In contrast estimating one dimensional laws is quite compatible with the amount of data.

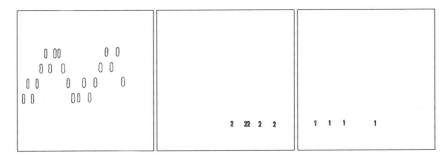

Fig. 7.11 Grouping with respect to orientation, elongation, diameter, and a convexity coefficient. The DNA markers are the most meaningful group $NFA_g = 10^{-10}$. Note that the 1 and 2's, though not meaningful groups, have NFA_g only slightly larger than 1 (1.6 and 4.5)

7.6 Bibliographic Notes

Finding groups in a large data set is an active research field. It is involved in data-mining, pattern recognition and pattern classification. The main clustering techniques are presented in [170, 59, 55, 96] and will be shortly reviewed in the keynotes

(Sect. A.1). Dubes [58] and Milligan and Cooper [125] proposed solutions to the choice of the number of clusters, which are related to the stopping rule in hierarchical methods. Bock [24] and Gordon [72, 73] are particularly interested in the validity assessment. Their approach is close to an *a contrario* method: they define a background model in which they measure the concentration of points. A uniform model may not be the best method, but it may be useful to define a data-dependent background model as done in the next chapter. The method of the present chapter is directly inspired by Desolneux *et al.* method for detecting dots in an image [53]. In this method, a hierarchical classification of the set of dots is considered and meaningful clusters are detected *a contrario* to a standard Poisson null model. A maximality criterion was also defined but had several flaws that were taken into consideration in this chapter. A very complete study of hierarchical segmentation and representation is presented by Guigues in his Ph.D thesis [77] (in French).

Chapter 8
Grouping Spatially Coherent Meaningful Matches

Structure means recognizing that unity is at the foundation of everything. To say structure is also to say Abstraction: geometry, rhythm, proportion, lines, planes, idea. These are elements of work – they act, they form, they construct and gain meaning through the law of unity.

Joaquín Torres-García, Estructura.

Abstract This chapter is about forming coherent groups of matching shape elements. This grouping, sometimes called generalized Hough transform, is crucial for virtually all shape recognition algorithms. It will be used for the main shape identification method treated in this book, namely the LLD, the MSER method in Chap. 9 and the SIFT method in Chap. 11. Each pair of matching shape elements leads to a unique transformation (similarity or affine map). A natural way to group these shape elements into larger shapes is to find clusters in the transformation space. The theory in the previous chapter is immediately applicable. The main problem addressed here is the correct definition and computation of the *background model* π. This background model is a probability distribution on the set of similarities, or on the set of affine transformations. In order to have accurate shape clusters, π must be built from empirical measurements on observable shape matching transformations. As in Chap. 5, the main issue is to compute accurately a density function in high dimension (4 or 6) with relatively few samples. The found solution is analogous: determine the marginal variables for which an independence assumption is sound. Then the density functions of these marginal laws can be accurately estimated on the data and yield an accurate background model.

8.1 Why Spatial Coherence Detection?

Figure 8.2 displays on the bottom left image a detail of Picasso's painting *Guernica* shown on the top left image (the original painting can be seen in Fig. 8.1). However, the painting is incomplete and partially occluded in the bottom image. It is also deformed by perspective. Moreover, the compression rates are also different. Figure 8.3 displays the LLDs common to these two images, both local and global, with affine invariant encoding. It turns out that local LLDs are much more discriminative. Indeed, since no restriction is made on the affine distortion, a lot of normalized convex LLDs look quite the same. The matching pairs have been computed by the method of Chap. 5. There are 94, whereas more global matches are due to quasi convex LLDs.

F. Cao et al., *A Theory of Shape Identification*. Lecture Notes in Mathematics 1948.
© Springer-Verlag Berlin Heidelberg 2008

Fig. 8.1 Pablo Picasso's *Guernica*. The choice of this masterpiece is related to Picasso's technique in this particular painting, that uses well contrasted strokes very accessible to a computer analysis (see Fig. 8.2). ©Succession Picasso/VG Bild-Kunst, Bonn 2008

Fig. 8.2 Guernica experiment. Original images and maximal meaningful level lines. Top: query image. Bottom: scene image. ©Succession Picasso/VG Bild-Kunst, Bonn 2008

The objective of this chapter is twofold: first, to prove that shape elements corresponding to a single shape can be accurately grouped together. Second, that this grouping procedure is robust enough to discard all false matches. Incidentally, this will dramatically reinforce the confidence in the more local previous detections. The group NFAs are indeed usually very small.

Fig. 8.3 Guernica experiment: meaningful matches, both affine invariant semi-local and global encoding. The number of semi-local LLDs is 7440 in the first image and 6131 in the second one (hence $4.6 \, 10^7$ tests). The number of globally encoded LLDs is 740 (resp. 897). There are very few false matches for locally encoded shapes and their NFA is more than 0.4. The total number of local (resp. global) matches is 94. (resp. 337). Globally encoded shapes yield many wrong affine invariant matches that can be observed on the left hand part of the second image. This is easily explained. All parallelograms are affine equivalent, and so are all triangles, or all ellipses. ©Succession Picasso/VG Bild-Kunst, Bonn 2008

The organization of this chapter is as follows. In Sect. 8.2, the parameterization of similarities or general affine transformations is described. Section 8.3 applies the general clustering ideas presented in Chap. 7 first by defining a dissimilarity measure between transformations, and then by defining a suitable background model for the sets of transformations. A few experiments are also shown to illustrate the ideas. Many more results will be given in the next chapter.

8.2 Describing Transformations

Let $\mathcal{I}$ and $\mathcal{I}'$ be two images, referred to as the *query* image and the *scene* image. For each match between a shape element S in $\mathcal{I}$ and a shape element S' in $\mathcal{I}'$, a geometric transformation (a similarity or an affine transformation) can be computed. In what follows, the parameters involved in these transformations are described as well as the way they can be estimated, both for the similarity and the affine transformation cases.

8.2.1 The Similarity Case

This subsection specifies the grouping method for the LLD case. Of course any other kind of shape element can be used provided matches yield candidate similarities or candidate affine transformations. Let S and S' be two matching LLDs. Recall that a LLD is a normalized piece of level line described in a local frame. It is completely

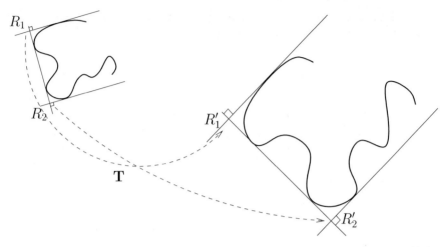

Fig. 8.4 Two pieces of level lines and their corresponding local similarity frames. The similarity $\mathbf{T}$ maps R_1 into R'_1 and R_2 into R'_2

determined by two points, or equivalently a point and a vector. This last representation will be chosen. A local frame is then given by a couple (p, v) where p gives the origin of the frame and v gives its scale and orientation. Assume that $\mathcal{S}$ is related to (p, v) and $\mathcal{S}'$ to (p', v'). Since $\mathcal{S}$ and $\mathcal{S}'$ match, they differ by a similarity transformation. Now, there exists a unique similarity mapping the local frame (p, v) onto (p', v') (see Fig. 8.4). By using complex numbers notation, this similarity can be uniquely expressed as

$$\forall z \in \mathbb{C}, \; \mathbf{T}(z) = az + b, \text{ with } a = \frac{v'}{v} \text{ and } b = p' - ap, \tag{8.1}$$

with $(a, b) \in \mathbb{C}^2$.

8.2.2 The Affine Transformation Case

Consider the local affine invariant normalization described in Chap. 4. Affine normalization of a piece of curve was done by mapping its local frame $\{R_1, R_2, R_3\}$ onto the triplet $\{(0, 0), (1, 0), (0, 1)\}$. Given another triplet $\{R'_1, R'_2, R'_3\}$ of non aligned points, there is a unique affine transformation mapping $\{R_1, R_2, R_3\}$ on $\{R'_1, R'_2, R'_3\}$, again denoted by $\mathbf{T}$. There exists a unique 2×2 matrix $\mathbf{M}$ and a unique $(t_x, t_y) \in \mathbb{R}^2$ such that

$$\mathbf{T}(x, y) = \mathbf{M} \begin{pmatrix} x \\ y \end{pmatrix} + \begin{pmatrix} t_x \\ t_y \end{pmatrix}$$

Calculating $\mathbf{M}$ boils down to the solution of a 2×2 linear system. By the classical QR decomposition [71], $\mathbf{M}$ can be written

$$\mathbf{M} = \begin{pmatrix} \cos\theta & -\sin\theta \\ \sin\theta & \cos\theta \end{pmatrix} \begin{pmatrix} 1 & \varphi \\ 0 & 1 \end{pmatrix} \begin{pmatrix} s_x & 0 \\ 0 & s_y \end{pmatrix}. \tag{8.2}$$

This decomposition is unique and completely determines $(\theta, \varphi, s_x, s_y)$ in $[0, 2\pi) \times \mathbb{R} \times \mathbb{R}_+ \times \mathbb{R}_+$. Let us denote by (x_{R_1}, y_{R_1}) and by (x'_{R_1}, y'_{R_1}) the pair of coordinates of R_1 and R'_1 respectively and by (m_{ij}) the coefficients of $\mathbf{M}$. Then the transformation parameters $T = (\theta, \varphi, s_x, s_y, t_x, t_y)$ can be computed by means of the following formulas

$$\begin{aligned} \theta &= \arctan(m_{21}/m_{22}), \\ \varphi &= (m_{11}m_{12} + m_{21}m_{22}) / (m_{11}m_{22} - m_{12}m_{21}), \\ s_x &= \sqrt{m_{11}^2 + m_{21}^2}, \\ s_y &= (m_{11}m_{22} - m_{12}m_{21}) / \sqrt{m_{11}^2 + m_{21}^2}, \\ \begin{pmatrix} t_x \\ t_y \end{pmatrix} &= \begin{pmatrix} x'_{R_1} \\ y'_{R_1} \end{pmatrix} - M \begin{pmatrix} x_{R_1} \\ y_{R_1} \end{pmatrix}. \end{aligned} \tag{8.3}$$

The vector T characterizes the transformation $\mathbf{T}$. Unambiguously one can adopt the same notation for similarities or affine transformations. In addition, since T characterizes $\mathbf{T}$, both of them can be identified. Thus write for $X \in \mathbb{R}^2$, $T(X)$ instead of $\mathbf{T}(X)$.

Figure 8.5 shows three 2-D projections of the transformation points T_k corresponding to the Guernica affine invariant meaningful matches (Fig. 8.3).

8.3 Meaningful Transformation Clusters

The problem of planar shape detection is by now reduced to a clustering problem in the transformation space. According to Chap. 7, it is necessary to define

1. A dissimilarity measure between points in the transformation space;
2. A probability on the space of transformations;
3. A grouping strategy.

8.3.1 Measuring Transformation Dissimilarity

Defining a distance between transformations is not trivial for two reasons. First, the magnitudes of the parameters of a transformation are not directly comparable. This

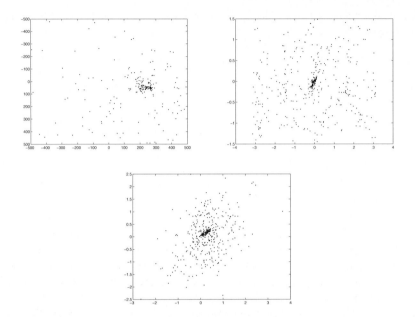

Fig. 8.5 Guernica experiment: each point represents a transformation associated with an affine invariant meaningful match, described by 6 parameters. Each figure represents a two-dimensional projection of the points, respectively t_x vs. t_y (translation coordinates), θ (rotation) vs. φ (shear), and $\ln(s_x)$ vs. $\ln(s_y)$ (zooms in the x and y directions). The noise is mainly due to similar global LLDs, which do not belong to the same real shape. The main cluster is also spread out because of the perspective effect. ©Succession Picasso/VG Bild-Kunst, Bonn 2008

problem is not specific to transformation clustering but general to clustering of any kind of data, as discussed in Sect. A.1. Second, our representation of similarities or affine transformations does not behave well in a vector space. A sound distance is not necessarily derived from a norm.

Definition 17 (similarity case). Let T (resp. T') be the similarity determined by two shapes elements $(\mathcal{S}_1, \mathcal{S}_2)$ (resp. $(\mathcal{S}'_1, \mathcal{S}'_2)$). Let also (R_1, R_2) (resp. (R'_1, R'_2)) be the points determining the local frame of $\mathcal{S}_1$ (resp. $\mathcal{S}'_1$). We call dissimilarity measure of T, T',

$$d_S(T, T') = \max \left\{ \|T(R_i) - T'(R_i)\|, \ \|T(R'_i) - T'(R'_i)\|, \ i \in \{1, 2\} \right\}. \quad (8.4)$$

Lemma 8. *The function d_S is non negative, symmetric and satisfies*

$$d_S(T, T') = 0 \Leftrightarrow T = T'.$$

Proof. The first two properties are obvious. Since a similarity is uniquely defined by the images of two points, the last property follows. Remark that d_S is not a distance since the triangle inequality does not hold. □

For the sake of completeness, let us define a dissimilarity function between affine transformations.

Definition 18 (affine case). Let T (resp. T') be an affine transformation determined by two shapes elements (S_1, S_2) (resp. (S_1', S_2')). Let also (R_1, R_2, R_3) (resp. (R_1', R_2', R_3')) the points determining the local frame of S_1 (resp. S_1'). We set

$$d_A(T, T') = \max\left\{ \|T(R_i) - T'(R_i)\|, \ \|T(R_i') - T'(R_i')\|, \ i \in \{1, 2, 3\} \right\}.$$
(8.5)

8.3.2 Background Model: the Similarity Case

In order to apply the detection framework of Chap. 7, a background law is first needed. A data point here is a similarity transformation represented by a pair of complex numbers $(a, b) \in \mathbb{C}^2$. The purpose of this section is to devise a sound background law π on the set of similarity transformations. With this in aim, recall that (a, b) is determined by two local frames in the images to be matched, respectively (p, v) and (p', v'). Let us now assume that these observations are the realization of a random variable $(P, V, P', V') \in \mathbb{C}^4$. It is natural to assume that the position, the size and the orientation of an object are independent. This is certainly sound up to some border effects. In addition, two images which do not contain common shapes also can be assumed independent. This leads us to take to the following independence assumption for the background model.

(A') *Consider a random model image $\mathcal{I}$ and a random scene image $\mathcal{I}'$. Then the random variables $P, |V|, \arg V, P', |V'|, \arg V'$ associated with (necessary casual) matches between both images are mutually independent.*

The marginal laws of the six previous random variables are easily learned from the two images. Hence the law of (P, V, P', V') is assumed to be known. By (8.1) such a 4-tuple uniquely defines a random similarity pattern denoted by (A, B) where A represents the rotation and zoom, and B the translation. The background law π is nothing but the distribution of (A, B). The expression of (A, B) as a function of (P, V, P', V') is explicit and given by

$$(A, B) : (P, V, P', V') \mapsto \left(\frac{V'}{V}, P' - \frac{V'}{V} P \right).$$

The background law π is the image of the law (P, V, P', V') by this application. It is also clear that A and B are not independent. Nevertheless, by definition of the conditional law,

$$d\pi(a, b) = d\pi^B(b \,|\, A = a) \, d\pi^A(a),$$
(8.6)

where π^A is the marginal of A and $\pi^B(\cdot \,|\, A = a)$ is the law of B knowing $A = a$. Since $|A| = |V'|/|V|$ and $\arg A = \arg V' - \arg V \mod (2\pi)$, these two variables are independent under Assumption **(A')**. Thus, the distribution π^A can easily be computed. Moreover, it turns out that A is independent from P and P'. Hence, the law of $B = P' - AP$, conditionally to $A = a$ is the law of $P' - aP$, which can also be easily computed under **(A')**. The background law π follows from (8.6).

In practice, the computation of π between two images is as follows:

1. Extract all the shape elements of query and scene images;
2. Compute the empirical laws of P, V, P', V' giving the position, the scale and the orientation of the local frames related to shape elements in the two images. Under the independence assumption **(A')**, this yields the law of the background model (P, V, P', V');
3. Under the same assumption, compute the empirical laws of $|A| = \frac{|V'|}{|V|}$ and $\arg A = \arg V' - \arg V \mod (2\pi)$;
4. For each value a of A with non null frequency, compute the empirical distribution of $P' - aP$.

The probability of a region R is then given by approximating the integral

$$\pi(R) = \int_R d\pi^B(b|A = a)\, d\pi^A(a).$$

A few words about the estimation of the background model: One would expect $\arg A$ to be uniformly distributed in $[-\pi, \pi)$, and this belief was experimentally confirmed, although the horizontal and vertical directions may sometimes be privileged. (See Fig. 8.6 and experiments.) The distribution of the zoom factor $|A|$ is instead far from being uniform or even showing a constant shape in the different experiments we have made. There is no way to figure out a realistic *a priori* distribution for $|A|$, or for B given A. The background model distributions must be learned from the scene and query images.

Remark 6. The ideas presented here also hold for the affine transformation clustering. For this case, θ, φ, s_x and s_y are considered to be mutually independent. Their distributions can be learned empirically as well as the joint probability of (t_x, t_y) given $(\theta, \varphi, s_x, s_y)$. This construction, experimentally satisfying though it is (see next chapter), has no right theoretical justification. The problem of finding the right independent marginal variables in the affine case is left open.

8.4 Experiments

The consistency of the previous definitions is now briefly checked. The next chapter contains many more experiments. All the experiments will be performed with a pair of images. The steps leading to a complete experimental setting for shape identification are:

1. The method of Chap. 5 is first applied and yields a set of M pairs of matching LLDs, one in the query image and one in the scene image;

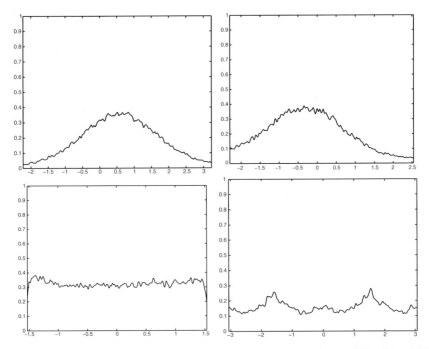

Fig. 8.6 Empirical histograms for affine invariant matching for the experiment of Figs. 8.2 and 8.3. In the first row, the empirical zoom factors in the x and y direction (logscale), which are image dependent. In the second row, the distribution of the shear and the rotation angle. The shear is basically uniform, but the rotation exhibits some peaks around $-\frac{\pi}{2}$ and $\frac{\pi}{2}$ because of the numerous horizontal and vertical lines in the image

2. A background model π on the set of similarities or on the set of affine transformations E is built using the method of the present chapter;
3. The transformations $T_1, \ldots, T_M$ associated with the matching pairs are used to form a point data set in E. From this set, a clustering tree is built according to the dissimilarity measures of Def. 17 (similarity case) or Def. 18 (affine case).
4. Maximal groups are computed by Def. 16.

The final outcome of the shape identification method of this book for each pair of images is a set of maximal meaningful clusters. Each cluster is likely to correspond to an identified shape. One can display for each cluster its associated LLDs. If the grouping is correct, this set of LLDs must correspond to a *matching shape* in both the query image and the scene image. In practice, the identified shapes have dramatically low NFAs. Thus, they yield an overwhelming certainty about identification. This certainty is, however, not fully unambiguous because of a stroboscopic effect. Indeed shapes often have self-similar parts: windows, or rows of windows in a building are a good example. Other examples are produced by symmetric shapes. For instance, the letter N is self-similar by a π rotation. In some cases, two or more very meaningful groups can be found, each one corresponding to a shape self-similarity.

Fig. 8.7 Guernica experiment: a single maximal meaningful group was detected. Matches of the group for the query image (left) and the scene image (right). The group is composed of 117 good matches and its $-\log_{10}(\text{NFA}_g)$ is 196.2. ©Succession Picasso/VG Bild-Kunst, Bonn 2008

Such self-similarities can, however, easily be anticipated by a previous comparison of the query image with itself. This comparison can be performed by the above algorithm. The main group will then correspond to the global match of the shape with itself and the other groups to stroboscopic effects.

Figure 8.7 depicts the maximal meaningful groups for the Guernica experiments. There is one single maximal meaningful group, with $-\log_{10}(\text{NFA}_g) = 196.2$. The best match between shape elements has a NFA about $4.16 \ 10^{-12}$. Hence grouping dramatically increases confidence in detections while all the false matches are eliminated. Marginals of the estimated background law were already shown in Fig. 8.6. This figure shows the learned distribution of the zoom factors in the x and y directions as well as the shear and rotation angle. The latter is not perfectly uniform in this case, because the vertical and horizontal directions are privileged in these geometrical images. Figure 8.8 shows the maximal meaningful cluster.

Now we consider the problem of finding common groups of shapes between the pair of images in Fig. 8.9. The same procedure is applied, in its similarity-invariant version. Two maximal meaningful groups are detected: the faces and the title. The corresponding points in the similarity space are displayed on Fig. 8.10. The two groups with their different translation and their different scaling are clearly visible this time.

The clustering algorithm decides that two separate groups (the actors' faces on the one hand and the word "Casablanca" on the other hand) are a better representation than a single large group containing both groups. Indeed, the large group in Fig. 8.11 has a NFA_g of $10^{-31.9}$, which is larger than the NFA_g of one of its children (whose values are $10^{-32.85}$ and $10^{-17.62}$). It follows from Def. 16 (Chap. 7) that the large group cannot be maximal. Still, one could argue that, since the values $10^{-31.9}$ and $10^{-32.85}$ are extremely close, the robustness in this decision does not seem to be in agreement with the situation depicted in Fig. 8.10, were the separation between the two detected maximal meaningful clusters seems to be quite large. Actually, the strong condition in this example, that prevents the large group from being maximal, is indivisibility. Indeed, by comparing the product of the children's NFA_g

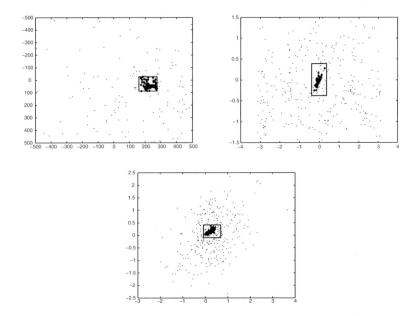

Fig. 8.8 Guernica experiment: transformation points of Fig. 8.5, where the points corresponding to the only affine invariant group are represented with larger dots. The boundaries of the corresponding hyperrectangle are drawn. The rest of the points are considered to be isolated and do not belong to any group. ©Succession Picasso/VG Bild-Kunst, Bonn 2008

with the large group's NFA_g, Prop. 11 (Chap. 7) is sufficient to state that the latter one is not indivisible, by far, and thus cannot be maximal.

The examination of the transformation histograms (Fig. 8.12) shows that the rotation angle is nearly uniformly distributed. The zoom factor, on the other hand, does not have an intuitive distribution. The translation has to be learned conditionally to the rotation and the zoom. The last two plots are the two-dimensional distribution of the translation, conditioned by the rotation and zoom of the two detected maximal meaningful groups. As can be seen, these distributions are not simple and cannot be deduced from one another by a single scaling.

8.5 Bibliographic Notes

The use of spatial coherence for shape or object detection has been the subject of intensive research, in particular since Ballard's work on the generalized Hough transform [14]. In his paper, Ballard proposed a method extending the Hough transform to any kind of planar shape not necessarily described by an analytic formula. Stockman [169] presented another early work based on the same principle (recognizing a query shape by finding clusters in the transformation space) where he introduced a

12 meaningful matches, $-\log_{10}(\mathrm{NFA}_g) = 32.85$

8 meaningful matches, $-\log_{10}(\mathrm{NFA}_g) = 17.62$

Fig. 8.9 Casablanca experiment: there are exactly two maximal meaningful groups, corresponding to the faces and the title. The relative scale of the images presented above is the same as the original one. One should note that the faces and the title actually lie in different relative positions and scales

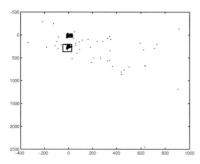
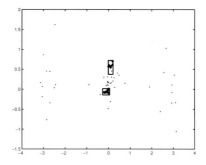

Fig. 8.10 Casablanca experiment: meaningful clusters in the similarity space. Left: projection on the translation dimensions. Right: projection on the rotation and zoom (log scale) axes. In this case, two clusters are clearly visible. Their position but also their scale are different

coarse to fine technique reducing the search complexity. Other voting schemes, like Geometric Hashing [184, 102] or the Alignment method [89], are frequently used

Fig. 8.11 Casablanca experiment. Meaningful group corresponding to the merging of groups in Fig. 8.9. This group contains 20 meaningful matches, and its $-\log_{10}(\text{NFA}_g)$ is 31.9. According to Definition 15 and following Prop. 11, it is not indivisible and cannot be maximal

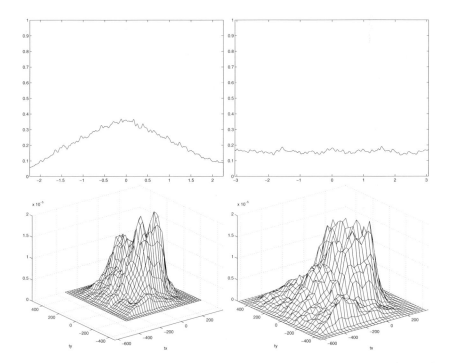

Fig. 8.12 Empirical histograms for similarity invariant matching for the experiment in Fig. 8.9. On the first row the log-empirical zoom factor $\ln(s)$ and the rotation angle θ. This last one is nearly uniform in this case. On the bottom row the distribution of the translation vector, conditioned by two different values of the couple $(\ln(s), \theta)$. These values correspond to the two maximal groups that are depicted on Fig. 8.9. Since the scales are different, so are the distributions

for detection or recognition. They are computationally more expensive and can be less accurate. In [75, 76] Grimson and Huttenlocher presented a study on the likelihood of false peaks in the Hough parameter space. Their work inspired the detection method adopted in this chapter. They indeed proposed a detection framework where recognition thresholds are derived from a null model (*the conspiracy of random-ness*). Previous recognition methods generally associated a single threshold with each query image, independent of the scene complexity. In contrast to these meth-ods, the grouping thresholds derived in this chapter satisfy an important property: they are functions of the scene complexity and of the uncertainty in feature extrac-tion. The method of the present chapter took these fundamental ideas from Grimson and Huttenlocher's work. The computational swiftness is obtained by hierarchically representing the transformation points. The definition of a data-dependent back-ground model is crucial for avoiding false clusters: Grimson and Huttenlocher's method assumes that matched features are uniformly distributed in the image. This assumption is usually not valid. See [147].

Finding groups in data sets is a major problem in many fields such as statis-tical pattern recognition, image processing, or data mining. Grouping phenomena are probably essential in human perception. In vision, the grouping phenomenon was thoroughly explored by the Gestalt school. The seminal paper on this problem is Wertheimer [179] . In Computer Vision, the first attempts to model a computa-tional perceptual organization date back to Marr [116]. More recently Lowe [112] proposed a detection framework based on computing accidental occurrences. He writes:

> In other words, one can shift our attention from finding properties with high prior expecta-tions to those that are sufficiently constrained to be detectable among a realistic distribution of accidentals.[...] Even when we ignore the ultimate interpretation for some grouping and therefore its particular a priori expectation, we can judge it to be significant based on the non-accidentalness criteria.

Lindenbaum's beautiful paper [106] proposed to evaluate *a priori* the performance of any invariant shape recognition device. For this author, a shape can be distin-guished if and only it occurs with very small probability in the random background. The author gives a lower bound on the number of points k in the shape ensuring recognition. This lower bound depends on the number of points N in the back-ground, the accuracy d of the recognition, the required invariance and ε, the allowed error probability for each test. Unfortunately, the shape model assumed by the au-thor is not quite realistic. For him, a shape is a cluster of points concentrated around some curve representing the shape's boundary and the background is modelled as a Poisson noise with lower density.

In [23], the authors have proposed a probabilistic *compositional model* for shape recognition. In their vision model, visual primitives are recursively composed, sub-ject to syntactic restrictions, to form tree-structured objects. The involved compo-sitional rules have a structure close to Chomsky's grammars. To take a simple but significant example of syntactic rule:

$$alignment + alignment \rightarrow alignment,$$

with the restriction that both alignments are themselves aligned. From the proba-bilistic viewpoint, the source of inspiration of this theory is very close to our own aims. This is best illustrated by the following quotation from Laplace's Essay on Probability which we take from [23].

> *On a table we see letters arranged in this order,* Constantinople, *and we judge that this arrangement is not the result of chance, not because it is less possible than the others, for if this word were not employed in any language we should not suspect it came from any particular cause, but this word being in use amongst us, it is incomparably more probable that some person has thus arranged the aforesaid letters than this arrangement is due to chance.*

Laplace is assuming *a contrario* that any combination of the 26 alphabet letters would be equally likely. Now, a modern dictionary contains not more than 10^5 words. The number of possible words with 14 letters like Constantinople in the *a contrario* model is about $2.4 \, 10^{19}$. Thus the probability of the group Constantinople happening just by chance is less than 10^{-14} in the *a contrario* model.

Chapter 9
Experimental Results

Abstract In this chapter we illustrate a complete and parameterless recognition process by presenting many experiments with manifold image kinds. The whole algorithm is concisely described in Appendix B.1.

9.1 Visualizing the Results

Almost all the experiments presented in this chapter are illustrated in a uniform format. The list below gives the format and summarizes the steps of the complete recognition algorithm.

1. *The two original images.*
2. *The smoothed maximal meaningful boundaries of the original images*, extracted using the algorithm described in Chap. 2, then smoothed with Moisan's implementation of the affine curve shortening equation (Chap. 3.3).
3. *Detection of meaningful matches between LLDs.* We consider here the 1-meaningful matches, despite the fact that a few of them may correspond to false detections. Indeed, as seen in Chap. 5, the constraints imposed by the encoding methods and by the non-intersection of level lines introduce a certain amount of dependence between the distances used as features in the *background model* (which were assumed to be independent). Thresholding the NFA at 0.1 empirically ensures that no detection occurs in white noise images. However, since the detection of meaningful matches is followed by a grouping process based on spatial coherence, in the experiments these few false matches are kept in order to test the robustness of the grouping algorithm.
   A fundamental hypothesis for the *a contrario* detection of groups is that under the *background model* transformation points are mutually independent. In order to comply with this hypothesis, a greedy algorithm eliminates matched LLDs which share a large piece of curve with other LLDs presenting lower NFA. More precisely if a pair of LLDs (S_1, S_1') is an ε_1-meaningful match, and there exists another pair (S_2, S_2') matching ε_2-meaningfully, with $\varepsilon_2 < \varepsilon_1$, such

that S_1 shares at least half of its length with S_2, and the same for S_1' and S_2', then the pair (S_1, S_1') is eliminated from the output list of matches.
The detection of 1-meaningful matches is illustrated by superimposing the matched LLDs to the original images.

4. *Grouping of spatially coherent meaningful matches.* For each meaningful group of matches that is detected (the maximal 1-meaningful groups defined in Chap. 8), four images are shown.

- *The LLDs that match within a group are shown, superimposed to the original images.*
- Given the set of transformations corresponding to the matches within a group, the best affine or even projective transformation (in the least squares sense) that maps the LLDs in the query image to the ones in the scene image is computed. Then the query image is mapped using this transformation. *The superposition of the transformed query image and the scene image is presented.*
- All the pieces of meaningful level lines of the two registered images are then submitted to a visual check. To this purpose, let us fix two values l and d. Let C_1 and C_2 be two pieces of level lines with the same length l parameterized by the length parameter. If for all $s \in (0, l)$, $|C_1(s) - C_2(s)| < d$ then display C_1 and C_2.

9.2 Experiments

The detection framework and the algorithms presented in this book are completely general and can be applied to any kind of images. Besides the Guernica and Casablanca experiments, this section gives some examples of different kind and nature, all similarities (or affinities). All experiments were done using the single-linkage algorithm (see Keynotes A.1.3).

9.2.1 Multiple Occurrences of a Logo

This example illustrates the performance of the proposed methodology in detecting multiple groups in an image. Two images containing occurrences of the Coca-Cola logo are compared in Fig. 9.1. Figure 9.2 shows the meaningful matches, both locally and globally encoded with the affine invariant method. They lead to points in the 6-dimensional space clustered by the single linkage method. Maximal meaningful groups appear in three projections in Fig. 9.3. Five groups are detected and are all correct. The corresponding LLDs are displayed for each group in Fig. 9.4. Figures 9.5 and 9.6 show the registration results of the query image into the scene image, for each maximal meaningful group.

Fig. 9.1 Coca-Cola experiment: original images and maximal meaningful level lines. Top: query image. Bottom: scene image

Fig. 9.2 Coca-Cola experiment: meaningful matches with local encoding (top) and global encoding (bottom). Number of tests: $1.57 \; 10^7$ (590 LLDs in the query image, $26,620$ in the scene image). There are 133 meaningful local matches and 1,002 global matches. The best match has $\text{NFA} = 8.4 \; 10^{-12}$

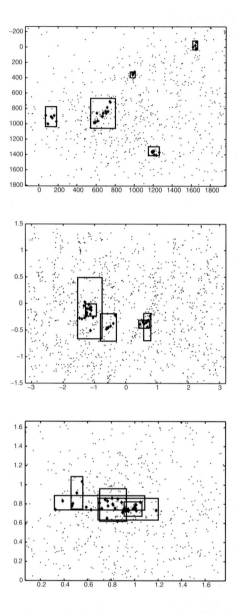

Fig. 9.3 Coca-Cola experiment: maximal meaningful groups and projections of the corresponding regions. Their $-\log_{10}(\text{NFA}_g)$ is respectively 19.4, 13.5, 3.7, 1.9 and 0.6. The first image corresponds to the projection on the (t_x, t_y) plane, where the groups are clearly separated. The second plot displays the rotation θ against φ (shear). Finally the last figure depicts the zoom in the x and y coordinates in the normalized frame (logarithmic scale). Identifying those groups in the point clouds is not easy. The points that belong to a maximal meaningful group are represented with larger dots. The rest of the points are considered to be isolated and do not belong to any group

Fig. 9.4 Coca-Cola experiment: the five maximal meaningful groups. Their $-\log_{10}(\text{NFA}_g)$ is respectively 19.4, 13.5, 3.7, 1.9 and 0.6, and the number of matches they contain is respectively 15, 7, 5, 6 and 4

Fig. 9.5 Coca-Cola experiment: superposition of the logo onto the image for the first three groups. A mean planar projective transformation is computed for each group of transformations by using a linear regression. Another practical way to check the validity of the transformation is to display all the pieces of maximum meaningful level lines that are everywhere close to each other after registration (in practice pieces of length 40 at distance less than 4)

Fig. 9.6 Coca-Cola experiment: superposition of the logo onto the image for the last two groups. See caption of Fig. 9.5

9.2.2 Valbonne Church

Figure 9.7 shows two different views of the Valbonne church with their corresponding maximal meaningful level lines. The meaningful matches between these two views up to similarity invariance are shown in Fig. 9.8. Some of them are false matches, but all of them showed a NFA larger than 0.1, as predicted by the experimental results in Chap. 5. There are also some casual matches that correspond to the same structures in the images. Figure 9.9 displays the only detected maximal meaningful group (see caption for details). A global affine transformation was estimated from this group by means of a least squares procedure, over the corresponding matched LLDs. This transformation was used to map the query image into the scene image (Fig. 9.10). The superposition of the transformed query image and the scene image shows that the estimated affine transformation is a good approximation to the actual projective transformation.

Fig. 9.7 Two frames of the Valbonne church sequence, with its corresponding meaningful level lines. The image on the top was taken as query image

Fig. 9.8 Valbonne church: 81 meaningful matches were found, for 14,710 LLDs in the query image and 18,413 in the scene image. All false detections have NFA larger than 0.1. The best match has NFA $= 2.98 \; 10^{-12}$

Fig. 9.9 Valbonne church: a single maximal meaningful group is detected. All false matches and spatially incoherent matches are rejected. The group contains 35 (similarity invariant) meaningful matches and $-\log_{10}(\text{NFA}_g) = 84.8$

Fig. 9.10 Valbonne church: registration of the images computed from the maximal meaningful group. On the right, the matching pieces of level lines. On the bottom right of the image some straight lines appear, because of the coincidental superposition of the lower and upper parts of the gate after registration

9.2.3 Tramway

The next experiment shows that the grouping procedure naturally induces a background/foreground separation. The two images to be matched, displayed in Fig. 9.11, are frames from a movie. Results are shown in Figs. 9.12 to 9.14. See captions for details.

Fig. 9.11 Tramway images: two frames from a movie

Fig. 9.12 Tramway images. Meaningful matches. 434 local and 71 global LLD matches

Fig. 9.13 Tramway images: the two maximal meaningful groups. The first one corresponding to the background contains 220 matches, and $-\log_{10}(\text{NFA}_g) = 643.7$. The second one is the train with 15 matches and $-\log_{10}(\text{NFA}_g) = 11.9$

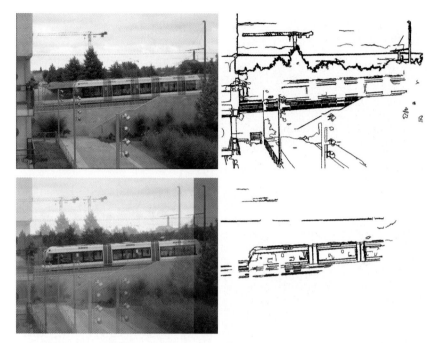

Fig. 9.14 Tramway images: registration with respect to two maximal meaningful groups. The first set of lines corresponds to the background. Notice the outer contour of the train. This is a consequence of the well-known aperture problem in optical flow computation: The visual motion cannot be determined in the direction of the level lines since this does not result in any change in the image. The counterpart is visible on the bottom images, where the motion of (static) cables cannot be separated from the motion of the tram

9.3 Occlusions

This section describes an example where the region of interest in the scene is occluded by the foreground. The images are two photographs of the painting *Las Meninas* by Velázquez. One is taken from the Web, and the other was shot directly in the *Museo del Prado* in Madrid. As can be seen in Fig. 9.15, the bottom image is partially occluded by people contemplating the painting. Colors and the illumination are completely different. Nonetheless, maximal meaningful level lines are quite insensitive to this change. This empirical statement will be proved by the matching and the grouping phase. The top image is the query and its shape content is sought for on the bottom image. In this experiment, the similarity version of the recognition method was used. Figure 9.16 shows the detected 1-meaningful matches between LLDs. The best match shows an NFA of $4.1\,10^{-14}$. Here again, the few false matches that were found have their NFA above 0.1.

Fig. 9.15 Las Meninas experiment. Top row: query image and its maximal meaningful boundaries. Bottom: scene image and maximal meaningful boundaries

Fig. 9.16 Las Meninas experiment: 1-meaningful matches. The NFA of the best match is $4.1\,10^{-14}$. Some false detections can be observed. All of them are due to global matches between nearly convex pieces

A single maximal meaningful group was detected. This group contains 70 spatially coherent meaningful matches, and its $-\log_{10}(\mathrm{NFA}_g)$ is 226.70. Figure 9.17 shows the matched LLDs that are within the group.

Fig. 9.17 The 70 matched LLDs within the spatially coherent group. All false matches have been rejected. The value of $-\log_{10}(\mathrm{NFA}_g)$ of the group is 226.70

The Las Meninas experiment shows as usual the superposition of the registered query image and the scene image (Fig. 9.18). The registration is very accurate, as can be seen in the pieces of level lines that are common to the two images. Nearly all the shape content is matched in this case.

Fig. 9.18 Las Meninas experiment. Left: superposition of the scene image and the transformed query image. Right: comparison of pieces of level lines

9.4 Stroboscopic Effect

In our last example, we will find groups of spatially coherent meaningful matches between the two images shown in Fig. 9.19. The meaningful matches between LLDs are displayed in Fig. 9.20. Three maximal meaningful groups are detected; their

NFA_g are reported in Tab. 9.1. The LLDs in each group and the registration results are shown in Fig. 9.21 and 9.22 . The most meaningful detection is of course overwhelming, with 169 matches and $-log_1 0(NFA_g) = 534.8$. The other two groups correspond to stroboscopic effects, and their detection is also very meaningful. These detections are completely legitimate since they correspond to large repeating parts of the image.

Fig. 9.19 Roundabout images

Fig. 9.20 Roundabout images and meaningful matches. There are 274 local matches, and 645 global ones

Table 9.1 Roundabout images. Number of matches and NFA_g of the meaningful groups (in the order depicted in Fig. 9.21)

Group nb.	1	2	3
nb. of matches	169	12	17
$-\log_{10}(NFA_g)$	534.8	76.81	46.2

Fig. 9.21 Roundabout images. There are three maximal meaningful groups. The NFA_g and the number of matches are reported in Tab. 9.1. All these groups are correct. The last two ones are due to the local self-similarity of the image

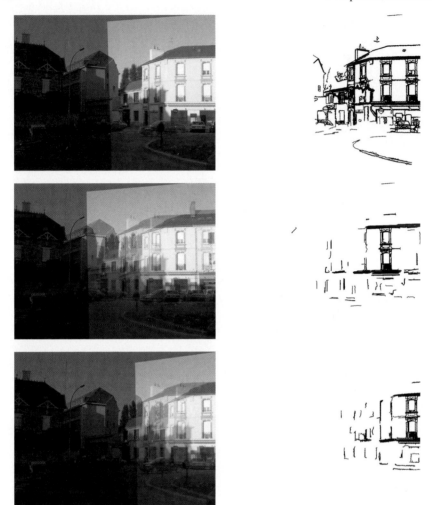

Fig. 9.22 Roundabout images. Left: superposition of the two images when the first one is mapped onto the second one by the planar projective mapping computed from maximal meaningful groups. The superposition of the two images is brighter in the overlapping area. On the right: pieces of level lines which coincide for both images

Part V
The SIFT Method

Chapter 10
The SIFT Method

Abstract In this chapter and in the next one, we describe one of the most popular shape descriptors, Lowe's *Scale-Invariant Feature Transform* (SIFT) method [114]. In continuation we will perform a structural and practical comparison of the SIFT-based matching method with the *Level Line Descriptor method* (LLD) developed in this book. The LLD method in fact includes the features of the recent, also popular, MSER method. Comparing SIFT and LLD is not an easy task, since they are of different nature. On the one hand LLD is based on geometrical shape descriptors, rigorously invariant with respect to similarity or affine transformations. Moreover, the method comes with decision rules, either for matching or grouping. On the other hand, SIFT descriptors are local patches which are based on key points and which are just similarity-invariant. The comparison will be based on *ad hoc* experimental protocols, in the spirit of the SIFT method itself. These protocols check the robustness of local descriptors to all perturbations listed in Sect. 1.2 (Chap. 1).

We start with a comprehensive description of the SIFT shape encoding (Sect. 10.1). Then we compare robustness and stability of both shape descriptors (Sect. 10.2). Sect. 10.3 compares the matching performances of both algorithms on pair of images having similar shapes or obtained by photographing the same scene under different viewpoints. The main focus of the book is the computation of matching thresholds. In the SIFT method the thresholds are learned from the pair of images. We shall see that obvious matches can be missed when the query shape appears more than once in the searched image. In the next chapter a fusion of SIFT and of the *a contrario* techniques both for matching and grouping will be proposed.

10.1 A Short Guide to SIFT Encoding

SIFT encoding is a procedure that enables one to extract local information from digital images. This procedure was introduced by Lowe in 2004 [114] and is now used in many computer vision applications. SIFT stands for Scale Invariant Feature Transformation: It consists in normalizing local patches around robust scale

F. Cao et al., *A Theory of Shape Identification*. Lecture Notes in Mathematics 1948.

covariant image key points. Lowe claims that 1) his descriptors are invariant with respect to translation, scale and rotation, and that 2) they provide a robust matching across a large range of affine distortions, change in 3D viewpoint, addition of noise, and change in illumination. In addition, being local, they are robust to occlusion. Thus they match all requirements for shape recognition algorithms listed in Sect. 1.2 (Chap. 1) except one: affine invariance.

This preliminary section is dedicated to the description of this encoding algorithm, which consists of four steps: detection of scale-space extrema (Sect. 10.1.1), accurate localization of key points (Sect. 10.1.2), orientation assignment (Sect. 10.1.3), and descriptor building (Sect. 10.1.4). We also briefly discuss the way SIFT descriptors can be compared (Sect. 10.1.5).

10.1.1 Scale-Space Extrema

Following a classical paradigm, stable points of interest are supposed to lie at extrema of the Laplacian of the image in the image scale-space representation. The scale-space representation introduces a smoothing parameter σ, the scale, and convolves the image with Gaussian functions of increasing standard deviation σ.

Thus digital images are smoothed at several scales: $L_\sigma = G_\sigma \star I$, where

$$G_\sigma = G(x, y, \sigma) = \frac{1}{2\pi\sigma^2} e^{-(x^2 + y^2)/2\sigma^2}$$

is the 2D-Gaussian function with integral 1 and standard deviation σ. The notation $\star$ stands for the convolution. By a classical approximation inspired from psychophysics [116], the Laplacian of the Gaussian is replaced by a Difference of Gaussians at different scales (DOG). Extrema of the Laplacian are then replaced by extrema of DOG functions: $D_\sigma = L_{k\sigma} - L_\sigma$, where k is a constant multiplicative factor. Indeed, it is easy to show that D_σ is an approximation of the scale-invariant Laplacian:

$$D_\sigma \approx (k - 1)\sigma^2 \Delta G_\sigma \star I.$$

In the terms of David Lowe:

The factor $(k-1)$ in the equation is constant over all scales and therefore does not influence extrema location. The approximation error will go to zero as k goes to 1, but in practice we have found that the approximation has almost no impact on the stability of extrema detection or localization for even significant differences in scale, such as $k = \sqrt{2}$.

To be more specific, one has

$$D(x, y, \sigma) = (G(x, y, k\sigma) - G(x, y, \sigma)) \star I(x, y) = L(x, y, k\sigma) - L(x, y, \sigma)$$

The relationship between D and $\sigma^2 \Delta G$ can be understood from the heat diffusion equation (parameterized in terms of σ rather than the more usual $t = \sigma^2$):

$$\frac{\partial G}{\partial \sigma} = \sigma \Delta G.$$

From this, we see that ΔG can be computed from the finite difference approximation to $\partial G/\partial \sigma$, using the difference of nearby scales at $k\sigma$ and σ:

$$\sigma \Delta G = \frac{\partial G}{\partial \sigma} \approx \frac{G(x,y,k\sigma) - G(x,y,\sigma)}{k\sigma - \sigma}$$

and therefore,

$$G(x,y,k\sigma) - G(x,y,\sigma) \approx (k-1)\sigma^2 \Delta G.$$

This shows that when the difference-of-Gaussian function has scales differing by a constant factor it already incorporates the σ^2 scale normalization required for the scale-invariant Laplacian.

This leads to an efficient computation of local extrema of D by exploring neighbourhoods through a Gaussian pyramid (see Figs. 10.1 and 10.2).

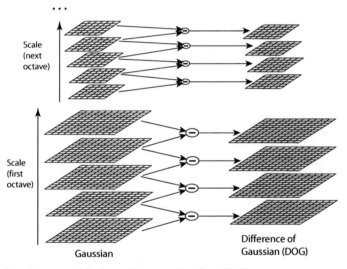

Fig. 10.1 Gaussian pyramid for key points extraction (from [114])

10.1.2 Accurate Key Point Detection

In order to achieve sub-pixel accuracy, the interest point position is slightly corrected thanks to a quadratic interpolation. Let us call $\mathbf{x}_0$ the current detected point in scale space, which is known up to the (rough) sampling accuracy in space and scale. Notice that all points $\mathbf{x}$ here are space coordinates supplemented with a scale coordinate. Let us call $\mathbf{x}_1 = \mathbf{x}_0 + \mathbf{y}$ the real extremum of the DOG function. Let us assume that $\mathbf{y}$ is small. The Taylor expansion of D_σ yields

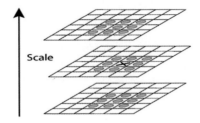

Fig. 10.2 Neighborhood for the location of key points (from [114]). Local extrema are detected by comparing each sample point in D_σ with its eight neighbors at scale σ and its nine neighbors in the scales above and below

$$D_\sigma(\mathbf{x}_0 + \mathbf{y}) = D_\sigma(\mathbf{x}_0) + \frac{\partial\,(D_\sigma(\mathbf{x}_0))^T}{\partial x}(\mathbf{x}_0)\mathbf{y} + \frac{1}{2}\mathbf{y}^T\frac{\partial^2 D_\sigma}{\partial x^2}(\mathbf{x}_0)\mathbf{y} + o(\|\mathbf{y}\|^2),$$

where D_σ and its derivatives are evaluated at an interest point and $\mathbf{x}$ denotes an offset from this point. Since interest points are extrema of D_σ, setting the derivative to zero gives:

$$\mathbf{y} = -\left(\frac{\partial^2 D_\sigma}{\partial x^2}\right)(\mathbf{x}_0)^{-1}\frac{\partial D_\sigma}{\partial x}(\mathbf{x}_0),$$

which is the sub-pixel correction for a more accurate position of the key point of interest.

Since points with low contrast are sensitive to noise, and since points that are poorly localized along an edge are not reliable, a filtering step is called for. Low contrast points are handled through a simple thresholding step. Edge points are swept out following the Harris and Stephen's interest points paradigm. Let H be the following Hessian matrix:

$$H = \begin{pmatrix} D_{xx} & D_{xy} \\ D_{xy} & D_{yy} \end{pmatrix}.$$

The reliability test is simply to assess whether the ratio between the largest eigenvalue and the smaller one is below a threshold r. It springs to check:

$$\frac{\text{Tr}(H)^2}{\text{Det}(H)} < \frac{(r+1)^2}{r}.$$

10.1.3 Orientation Assignment

Up to this point, key point extraction is scale-invariant. In order to extract rotation-invariant patches, an orientation must be assigned to each key point. Lowe proposes to estimate a semi-local average orientation for each key point. From each sample image L_σ, gradient magnitude and orientation is precomputed using a 2×2 scheme.

An orientation histogram is assigned to each key point by accumulating gradient orientations weighted by 1) the corresponding gradient magnitude and by 2) a Gaussian factor depending on the distance to the considered key point and on the scale. The precision of this histogram is 10 degrees. Peaks simply correspond to dominant directions of local gradients. Keypoints are created for each peak with similar magnitude, and the assigned orientation is refined by local quadratic interpolation of the histogram values.

10.1.4 Local Image Descriptor

Once a scale and an orientation are assigned to each key point, it is possible to extract similarity-invariant patches. The main problem is to extract *robust* patches.

Gradient samples are accumulated into orientation histograms summarizing the contents over 4×4 subregions surrounding the key point of interest. Each of the 16 subregions corresponds to a 8-orientations bins histogram, leading to a 128 element feature for each key point (see Fig. 10.3). Two modifications are made in order to reduce the effects of illumination changes: histogram values are thresholded to reduce importance of large gradients (in order to deal with a strong illumination change such as camera saturation), and feature vectors are normalized to unit length (making them invariant to affine changes in illumination).

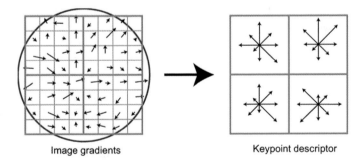

Image gradients   Keypoint descriptor

Fig. 10.3 Example of a 2×2 descriptor array of orientation histograms (right) computed from an 8×8 set of samples (left). The orientation histograms are quantized into 8 directions and the length of each arrow corresponds to the magnitude of the histogram entry. (From [114])

10.1.5 SIFT Descriptor Matching

Even for highly distinctive key point descriptors, Lowe admits that many false matches can be seen in the case of cluttered images. He proposes to address this problem by a Generalized Hough Transform step, in order to identify subsets of matching key points that also agree on location, scale, and orientation. He also proposes an alternative option, which consists in selecting the matches by thresholding the ratio between the distance from the query match to the first closest database match and the distance from the query match to the second one. This ensures the selected matches to be clearly separated from the clutter. An obvious drawback is that no repeated match can be detected.

The SIFT method is *a priori* perfectly similarity invariant and turns out to be, as we will see, robust to moderate affine transformations. Before passing to comparative experiments, let us point out that, according to Lowe, the elimination of outliers still is a problem. The SIFT method empirical threshold performs reasonably, but breaks down when several similar shapes are present (this is what we called the stroboscopic effect). Thus, we will demonstrate in the next chapter that an *a contrario* method can secure SIFT descriptors. Lowe recommends a grouping step, which is exactly what the grouping method developed in Chap. 8 does. Thus, this will also be developed in the next chapter.

10.2 Shape Element Stability versus SIFT Stability

10.2.1 An Experimental Protocol

The standard requirement for a shape matching algorithm is its invariance with respect to the classic image perturbations listed in Sect. 1.2, Chap. 1. In this section we shall compare the stability of the shape elements obtained by LLD to the stability of the SIFT descriptors. The most classical image perturbations will be considered: blur, contrast changes, viewpoint changes, similarity transformations, JPEG compression. The aim here is neither to compare the matching algorithms themselves, nor to estimate the accuracy of the descriptors (similarity invariant[1] curve descriptor for LLD or patch descriptor for SIFT). We just intend to control the stability of the *location* of these descriptors when the perturbations are applied. A first step in this experimental assessment is to define a common algorithm, so that it makes sense to compare the two families of descriptors. Considering a reference image (image 1) and five images (images 2 to 6) representing the same scene with some perturbation (change of viewpoint, illumination, blur, etc) and considering that the homographies H_i leading from image 1 to image i are known, we propose the following comparison protocol:

1. Extract descriptors from image 1 and image i.
2. Compute the proportion p of descriptors of image 1 that match, after transformation by H_i, with at least one descriptor of image i.

---

[1] Affine invariant LLDs are not considered in this comparison chapter since SIFT descriptors are just similarity invariant.

A similarity invariant image descriptor consists of a 2D location, an angle and a scale. When considering level line descriptors (LLDs), we take the two reference points in the canonical frame (similarity invariant encoding, see Sect. 4.2) and compute the magnitude, angle and midpoint of the segment joining them. SIFT descriptors are already associated with a point (the key point), an angle and a scale, from which a second point can be computed. The midpoint between these two points is used to indicate the 2D location of the descriptor.

Two descriptors are considered to match if their orientations, scales and midpoint locations are similar, up to some allowed error. In the tests we permit up to 5 pixels error in location, 80% change in scale and 30° angular variation. It is realistic to take such large thresholds, because the transformations between natural images are far from being perfect similarities and actually always have some affine distortion.

In order to prevent the possibility of counting several matches between the images corresponding to approximately the same spatial region, similar descriptors in image 1 (the ones that match according to the criteria specified above) are first grouped together. Therefore, p is actually the proportion of *groups of descriptors* in image 1 that find some match in image i.

Six experiments were led to test several invariance requirements. The test images are shown in Sect. 10.2.2 (let us recall that image 1 is the reference image), and the proportion p of descriptors which are retrieved is plotted in a graph.

Let us remark that if the descriptor extraction is robust and if both images have a similar spatial content, a high value of p must be expected. On the contrary, if this proportion is very low, any matching algorithm will fail, simply because the sought descriptors have not been extracted in the other image.

10.2.2 Experiments

The database used in this section comes from (see e.g. [121]):
http://www.robots.ox.ac.uk/~vgg/research/affine/index.html.

Each image comes with a homography matrix that permits to register each image from number 2 to number 6 to the first image of the series. (In Computer Vision, a planar homography is interpreted as a mapping between a point on a ground plane as seen from one camera, to the same point on the ground plane as seen from a second camera.) Extraction and matching of SIFT descriptors was performed by using Lowe's software from http://www.cs.ubc.ca/~lowe/keypoints/

10.2.2.1 Testing Robustness to Blur

Figure 10.4 displays a set of images obtained by an increasing real defocus of the camera. The results in Fig. 10.5-left show that SIFT descriptors behave better than LLD descriptors in the presence of blur. The reason is that SIFT key points are obtained as the result of a blurring process (which produces the scale-space). Indeed, the SIFT method actually *simulates the physical blur* at several scales, thus making the shape recognition, to some extent, blur invariant. On the contrary, level lines are highly perturbed by the blur and therefore LLDs greatly differ between the images.

Results for LLD can be improved by paralleling the SIFT method, namely by computing the LLDs for image 1 after applying a Gaussian blur with increasing variance (see Fig. 10.5-right).

Fig. 10.4 Images obtained by an increasing real defocus of the camera. Row 1: images 1 and 2. Row 2: images 3 and 4. Row 3: images 5 and 6

10.2.2.2 Testing Robustness to Zoom and Rotation Change

The deformation between images taken from a similar view point is mainly due to the zoom factor and the rotation of the camera around its optical axis. See Fig. 10.6.

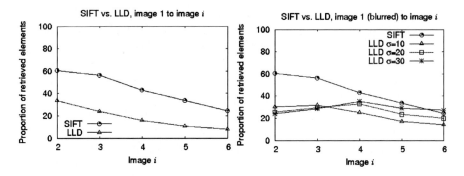

Fig. 10.5 Testing robustness to blur. Left: proportion of retrieved elements from image 1 to image *i*. Right: proportion of retrieved elements from image 1 + increasing Gaussian blur to image *i*

The results with both methods are quite similar (see Fig. 10.7), with a small advantage to LLD.

10.2.2.3 Testing Robustness to Viewpoint Change

The next experiment consists in retrieving the same graffiti from different viewpoints. Let us note that a strong perspective distortion can be seen in image 6 (Fig. 10.8).
The results here are slightly in favor of LLD (see Fig. 10.9). However, under strong perspective deformation none of the methods is able to produce good results.

10.2.2.4 Testing Robustness to Non-Linear Contrast Change

Simulating changes in illumination or in the non linear response of the CCDs, non-linear contrast changes and contrast inversions have been applied to a reference image, as can be seen in Fig. 10.10.
Results in Fig. 10.11 show that both methods are quite robust to contrast changes (somehow surprisingly SIFT descriptors are better than LLDs in one case). As expected, the SIFT robustness is very poor with respect to contrast inversion. The very few matches occur where descriptors with opposite direction and a similar scale have been extracted. As expected, LLDs behave very well since they are indeed designed to be robust with respect to any kind of contrast change and contrast inversion.

Fig. 10.6 Images undergoing different changes in rotation and scale. Row 1: images 1 and 2. Row 2: images 3 and 4. Row 3: images 5 and 6

10.2.2.5 Testing Robustness to JPEG Compression

In this experiment, the parameter of interest is the JPEG compression rate, which increases from image 1 to 6 (see Fig. 10.12).

LLD behaves poorly when compared to SIFT (see Fig. 10.13). The block effect destroys level lines because if the compression rate is too strong, level lines mostly follow blocks. This problem actually is, like the invariance to zooms, easily fixed by making the LLD method covariant to zooms. Indeed, with zooms applied the block effects are highly attenuated.

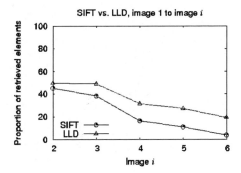

Fig. 10.7 Testing robustness to zoom and rotation change. Proportion of retrieved elements from image 1 to image i

10.2.2.6 Testing Robustness to Noise

In the next experiment images 2, 3, 4, 5 and 6 have been obtained by adding to image 1 increasing amounts of Gaussian noise (10, 20, 30, 40 and 50 standard deviation, respectively, see Fig. 10.14).

Here again SIFT descriptors outperform LLD (see Fig. 10.15-left). The reason is that noise destroys the level lines structure of an image and therefore alters its LLDs. On the contrary, Gaussian convolution reduces noise and SIFT descriptors are better preserved. As has been commented in Sect. 10.2.2.1, the use of a scale-space strategy in combination with LLD extraction brings back the performance of LLD to the level of SIFT (see Fig. 10.15-right).

10.2.3 Some Conclusions Concerning Stability

Although every comparison protocol is a questionable point, lessons can be learned from the preceding experiments and corresponding figures. We have compared two very different extraction procedures for shape elements, one based on level lines (LLD) and the other one on points of interest (SIFT). Both extraction algorithms exhibit a similar robustness with respect to contrast changes, similarities and weak viewpoint changes. LLD is unable to resist blur, JPEG compression and noise since they imply changes in the image geometry, while the blur invariant nature of SIFT key points (as stable points in the scale-space) descriptors makes them specially adapted to these kinds of perturbations. However, it has been shown that the use of a scale-space strategy dramatically increases the LLD performance. The main weakness of SIFT descriptors is their intrinsic modest robustness to strong similarities and viewpoint changes. On the contrary, LLDs can be made robust to affine transformations (see Chap. 6) which are local approximations to true perspective transformations.

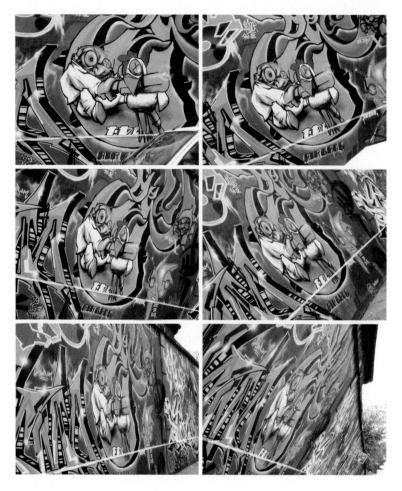

Fig. 10.8 Snapshots of the same graffiti from different viewing positions. Row 1: images 1 and 2. Row 2: images 3 and 4. Row 3: images 5 and 6

10.3 SIFT Descriptors Matching versus LLD *A Contrario* Matching

The previous section was about the descriptor repeatability and stability under several perturbations. In this section we compare the matching results of SIFT and LLD.

The LLD matching strategy has been extensively described in previous chapters and is mainly based on the use of *a contrario* techniques both for matching individual shapes and for grouping together matches contributing to similar homographies between the compared images.

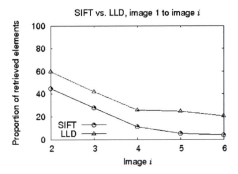

Fig. 10.9 Testing robustness to viewpoint change. Proportion of retrieved elements from image 1 to image i

Fig. 10.10 The same image after different contrast changes and contrast inversion. Row 1: images 1 and 2. Row 2: images 3 and 4. Row 3: images 5 and 6

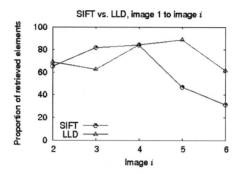

Fig. 10.11 Testing robustness to non-linear contrast change. Proportion of retrieved elements from image 1 to image i

In its simpler version, the SIFT matching is based on the following algorithm proposed by Lowe in [114]: A SIFT descriptor from the first image is matched with its nearest neighbor in the second one, provided the ratio of the distances between the nearest and the second nearest neighbor is below some threshold (this latest condition should reinforce the confidence, making this matching algorithm more stable than simply thresholding the distances). The lower the threshold is, the more reliable the matches should be. Despite this, one can observe in the experiments (see next section) that false matches are mixed with good matches, even with a decreased threshold (see [120] where this analysis is systematically led with several descriptors). This is not surprising since the ratio for SIFT is purely empirical and does not follow, as it should, from a probabilistic analysis. Some statistically rare descriptors would deserve a relaxed threshold, while more care should be taken with common descriptors. In order to get rid of the false matches mixed with good matches, a postprocessing based on the Generalized Hough Transform could be used (as proposed in [114]). However, voting thresholds and bin sizes are also touchy parameters that can introduce either false positives or wrong detections. For these reasons in our tests we have not used this postprocessing.

10.3.1 Measuring Matching Performance

Two magnitudes have been defined to quantify the quality of the matching between two images. First, if the homography between both images is known, we define the matching **efficiency** as the ratio between the number of correct matches and the total number of matches. A match between two pairs of descriptors is said to be correct if, after applying the known homography to the first element of the pair, the distance to the second element is small (in our tests, smaller than 5 pixels). In the case of SIFT, we use as matching elements the key points, and in LLD we use one of the points of the local reference frame of the shape.

Fig. 10.12 The same image with different JPEG compression rates. Row 1: images 1 and 2. Row 2: images 3 and 4. Row 3: images 5 and 6

Second, the **matching area** is the proportion domain of the first image that has a correct match in the second image. This is easy to compute when using SIFT, since each descriptor is associated to an image patch. However, LLD gives matches between pieces of level lines, not between bi-dimensional portions of the images. Thus, in order to compare both kinds of matches we have divided the domain of the first image into small squares with fixed size (50×50 pixels in our tests) and we consider that one of these squares has a match in the second image if it contains at least one correct matching element (key point for SIFT or piece of matched level line for LLD). The ratio between the number of matched squares and the total number of squares in the image is what we call area of the matching.

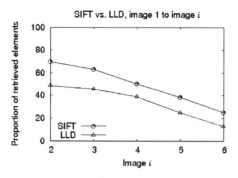

Fig. 10.13 Testing robustness to JPEG compression. Proportion of retrieved elements from image 1 to image i

Fig. 10.14 Images with increasing levels of white Gaussian noise. Row 1: images 1 and 2. Row 2: images 3 and 4. Row 3: images 5 and 6

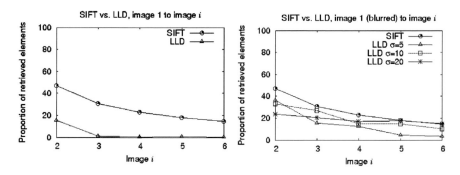

Fig. 10.15 Testing robustness to noise. Left: proportion of retrieved elements from image 1 to image i. Right: proportion of retrieved elements from image 1 + increasing Gaussian blur to image i

10.3.2 Experiments

10.3.2.1 Blur

We compared by SIFT and LLD the images in Fig. 10.4. Observe (Fig. 10.17) that the efficiency of both methods is high but that SIFT outperforms LLD in terms of matching area (see Fig.10.16). If we apply a blur to the first image, as suggested in section 10.2.2, matching results improve significantly for LLD (Fig. 10.17).

Fig. 10.16 Left: SIFT matching (white points correct matches, black points wrong matches). Right: LLD matching. The density of correctly matched SIFT descriptors is remarkable

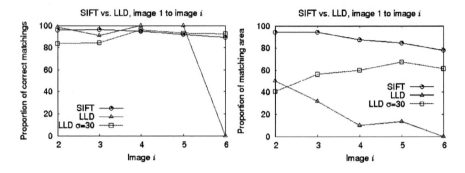

Fig. 10.17 Performance of matching with blur, between image 1 and image i. Left: matching efficiency. Right: matching area

10.3.2.2 Zoom and Rotation

We compare the images in Fig. 10.6. The results in Fig. 10.19 show that LLD and SIFT have similar performances under rotation changes. LLD performance decreases abruptly under strong scale changes (see Fig. 10.18).

10.3.2.3 Changing the Viewpoint

We compare the images in Fig. 10.8. First we use the similarity invariant encoding. The results in Fig. 10.22 show that SIFT, even if it is not, by design, affine invariant, is able to cope with affine deformations better than LLD in its similarity invariant version (see Fig. 10.20). Instead, LLD results are better when using the affine invariant version of the algorithm (see Fig. 10.21).

10.3.2.4 Stroboscopic Effect

The following figures display one of the main problems of the matching strategy for SIFT: it is unable to detect several instances of the same object (see Fig. 10.23 and 10.24). LLD is able to find as many groups of matches as instances of the same object are present in the image (see Fig. 10.25, 10.26, 10.27).

In the next chapter we will discuss some simple modifications of SIFT that enable the detection of several instances of the same object.

Fig. 10.18 Left: SIFT matching (white points correct matches, black points wrong matches). Right: LLD matching. Top: image 1; bottom: image 2

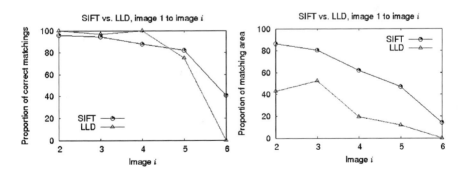

Fig. 10.19 Performance of matching with zoom and rotation, between image 1 and image i. Left: matching efficiency. Right: matching area

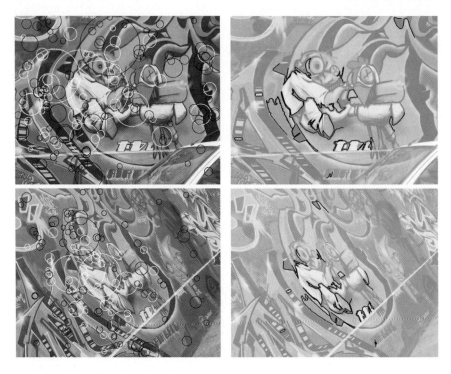

Fig. 10.20 Left: SIFT matching. Each circle represents a SIFT key point which got a match. The radius is the scale of the key point. White circles represent correct matches, black circles represent wrong matches. Right: LLD matching. Top: image 1; bottom: image 4

Fig. 10.21 LLD affine invariant matching. Left: image 1. Right: image 4

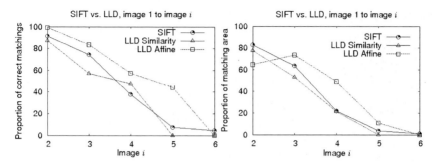

Fig. 10.22 Performance of matching with change in viewing position, between image 1 and image i. Left: matching efficiency. Right: matching area

Fig. 10.23 Original images. Left: image 1. Right: image 2

Fig. 10.24 SIFT matching. Left: from image 1 to image 2. Right: from image 2 to image 1

Fig. 10.25 LLD matching (first maximal meaningful group)

Fig. 10.26 LLD matching (second maximal meaningful group)

Fig. 10.27 LLD matching (third maximal meaningful group)

10.4 Conclusion

We have confirmed experimentally what could have been anticipated from the methods. SIFT, which simulates blur at various scales, is much more robust in presence of blur and noise. In order to put LLD at the same level, several blurs of the query images or shapes must be performed before LLD is applied to each one of them. Under weak viewing deformations SIFT also performs better than LLD in terms of matching area, since SIFT descriptors are more dense than LLD shapes. LLD is instead obviously more robust to strong affine deformations.

As expected, since it attempts to make accurate rejection thresholds, the efficiency of LLD is in general higher than SIFT in terms of rejection of wrong matches. The next chapter will consider the fusion of both methods, trying to take from each method what it is best for.

10.5 Bibliographic Notes

10.5.1 Interest Points of an Image

SIFT interest points (the key points) are obtained as the maxima of the Laplacian of the image (approximated by a difference of Gaussians) through a Gaussian pyramid. Many variations exist on the computation of interest points, following the pioneering work of Harris and Stephens [80]. In particular, recent methods are affine invariant. In [122], an overview and a comparison between the main affine invariant region detectors is presented. One of the conclusions is that no method really outperforms all the other ones, although the highest score is obtained by the MSER detector [118].

10.5.2 Local Descriptors

SIFT descriptors are basically local histograms of the gradient direction, weighted by the gradient norm, in the vicinity of the key point. These histograms are invariant to rotations of the image domain and thresholding and normalization of image gradients is used in order to achieve some invariance to illumination changes. In the recent years, several other local descriptors have been proposed, incorporating further invariance to changes in viewing conditions. In particular, MSER [118] uses moment invariants to describe the vicinity of the interest points. This approach was also used by Monasse in [133]. A recent paper [120] aims at comparing the different descriptors. Performance is evaluated by examining the so-called ROC curves plotting the number of false positive detections as a function of false negative detections. While on one of the methods, the gradient location and orientation histograms (GLOH [120]) seems slightly better than the other ones, the difference (in particular with SIFT) is not that large. Let us remark that there are two ways to achieve geometrical invariance: either descriptors are computed in invariant regions, or they have a group invariance by themselves. For instance, in [19], skew and stretch are corrected in the neighborhood computation. An affine contrast change is first applied. Then, descriptors are rotation invariant gray level moments.

10.5.3 Matching and Grouping

The matching phase relies on the distance between descriptors. A distance that is commonly used between descriptors is the Mahalanobis distance. It is basically a L^2 norm in an orthogonal basis (not orthonormal) where coordinates may be assumed uncorrelated. The implicit assumption is that distribution of descriptors is Gaussian, and there is no reason why this would be true. Moreover, the values of this distance have no absolute meaning: it merely allows to rank match candidates. Hence, simple procedures, such as the thresholding of ratios between the best and second best matches in SIFT, are usually used. MSER [118] uses a voting procedure over the nearest measurements comparing a set of invariants that form the descriptors. In the next chapter some improvements on the SIFT descriptors definition and on the matching step are proposed, based on *a contrario* techniques.

The use of a grouping step improves the matching results. In SIFT, a Hough transform procedure [14] is proposed, but other methods [157] use greedy procedures based on RANSAC [64]. The clustering method proposed in Chap. 8 and 9 can also be used to group the matching results, as shown in the next chapter.

Chapter 11
Securing SIFT with *A Contrario* Techniques

Abstract In the previous chapter two shortcomings of Lowe's SIFT algorithm have
been pointed out, namely its low matching efficiency (ratio between the number of
correct matches and the total number of matches) and its inability to match several
instances of the same object. The grouping stage of the method also is widely empirical and requires some fix.

In this chapter we shall examine three easy improvements of the SIFT method, all
based on the *a contrario* techniques developed in the present book. They permit to
treat all raised issues. The first one (Sect. 11.1) is the direct application of the theory for *a contrario* grouping of transformations developed in Chap. 8. The second
one (Sect. 11.2) is the use of a background model for SIFT matches which prevents
the elimination of multiple matches. Finally Sect. 11.4 yields an efficient *a contrario* technique computing a NFA for each SIFT match. In summary, the aim is to
demonstrate that the whole SIFT algorithm can be secured and associated realistic
NFAs, as we did in Chap. 5 and 8 for the LLD method.

11.1 *A Contrario* Clustering of SIFT Matches

The problem of matching efficiency of the SIFT algorithm was already remarked
in [114] by D. Lowe. He proposed to address this problem by a generalized Hough
transform, in order to identify subsets of matching key points that also agree on
location, scale, and orientation. Quoting Lowe:

> *The correct matches can be filtered from the full set of matches by identifying subsets of
> keypoints that agree on the object and its location, scale, and orientation in the new im-
> age. The probability that several features will agree on these parameters by chance is much
> lower than the probability that any individual feature match will be in error. The determi-
> nation of these consistent clusters can be performed rapidly by using an efficient hash table
> implementation of the generalized Hough transform. Each cluster of 3 or more features
> that agree on an object and its pose is then subject to further detailed verification. First,
> a least-squared estimate is made for an affine approximation to the object pose. Any other
> image features consistent with this pose are identified, and outliers are discarded. Finally,*

F. Cao et al., *A Theory of Shape Identification*. Lecture Notes in Mathematics 1948.   209
© Springer-Verlag Berlin Heidelberg 2008

*a detailed computation is made of the probability that a particular set of features indicates
the presence of an object, given the accuracy of fit and number of probable false matches.
Object matches that pass all these tests can be identified as correct with high confidence.*

In this section we propose to develop this technique and to replace the Hough
transform by a clustering step identical to the one described in Chap. 8 and 9.

The location, scale and orientation of each one of the SIFT matching pairs can
be represented as a point in the space of similarity transformations. These points
can be grouped together using the technique of Chap. 8. The resulting meaningful
clusters correspond to sets of spatially coherent matches. Matches not belonging to
any meaningful cluster are rejected as wrong.

The following images illustrate the use of this clustering technique. In Fig. 11.2
the clustering method has been applied to a pair already shown in the previous chapter. Each circle represents a SIFT key point that got a match. As usual, the radius
represents the scale of the key point. White circles represent correct matches, black
circles represent wrong matches. The results for the original SIFT method are shown
on the left part of the image. On the right side, the new results using the clustering
technique are shown. The most meaningful cluster is displayed and the key points of
the SIFT descriptors contributing to the cluster are shown in white. The efficiency
of the matching procedure increases from 37.9% to 67.8%. Notice that efficiency
did not reach 100% because some imprecise (though not completely incoherent)
matches were included in the maximal meaningful cluster. For all of them, the location and scale are quite close to the one expected according to the underlying
homography, this explains why these matches were included in the cluster.

The second figure (Fig. 11.1) compares the results of the original SIFT algorithm (left) and those obtained after clustering (right). As can be observed, the final
number of matches decreases but the efficiency increases significantly (from 82.1%
to 100%).

In the next figures we show an example with several clusters of matches.
Figure 11.3 displays the result of the comparison of two of the images displayed
in the previous chapter (Fig. 10.8). Observe that some of the matches are wrong.
After applying the clustering step two clusters of matches are found (see Fig. 11.4).
All of the matches in these clusters are correct.

11.2 Using a Background Model for SIFT

The matching algorithm for SIFT used in the previous chapter consists in selecting
the matches by thresholding the ratio between the distance from the query match
to the first closest database match and the distance from the query match to the
second closest one. This ensures the selected matches to be clearly separated from
the clutter. The best threshold was determined empirically by Lowe by making a
statistics on 40,000 key points:

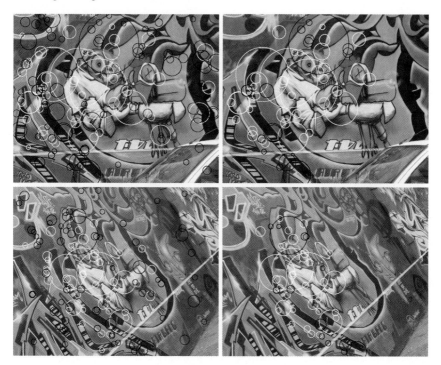

Fig. 11.1 Left: SIFT matching (the radius represents the scale of the key point; white circles represent correct matches, black circles represent wrong matches). Right: SIFT matching followed by *a contrario* clustering (white circles mark the key points belonging to the cluster, all of them are correct matches). Top: image 1; bottom, image 4

> *The probability that a match is correct can be determined by taking the ratio of distance from the closest neighbor to the distance of the second closest. Using a database of 40,000 keypoints (...).*

Actually it is clear that this empirical probability is not given by a model. Another obvious drawback is that no repeated match can be detected.

A very simple way to overcome this last drawback consists in using a third image as *background model* for learning the rejection thresholds. Here we are applying the very same SIFT method, but we just notice that any image can be used as background model (the one we have used in our tests is shown in Fig. 11.5). The matching procedure simply consists in selecting the matches by thresholding the ratio between the distance from the query match to the database match and the distance from the query match to the closest match in the background model. As a consequence of this modification several new matches have been detected and, in particular, several instances of the same object have now been detected (see Figs. 11.6 and 11.7).

Fig. 11.2 Left: SIFT matching (the radius represents the scale of the key point; white circles represent correct matches, black circles represent wrong matches). Right: SIFT matching followed by *a contrario* clustering. Top: image 1; bottom: image 5

Fig. 11.3 SIFT matching between top and bottom images

Fig. 11.4 SIFT matching between top image (query) and bottom image (database), followed by *a contrario* clustering. Two maximal meaningful clusters were detected

Fig. 11.5 Image used as background model for the rejection thresholds of matches

Fig. 11.6 Left: original SIFT matching result. Since the same shapes occur twice in the bottom image, the standard SIFT threshold procedure eliminates almost all matches. Right: result when using a background model. Top: query image; bottom: database image. The wrong matches can be eliminated by clustering in the transformations' space, as illustrated in the next figure 11.7

Fig. 11.7 SIFT matching between the top image (query) and the bottom image (database), with background model thresholding, followed by *a contrario* clustering. Two maximal meaningful groups were detected

11.3 Meaningful SIFT Matching

The SIFT algorithm is based on the use of a set of descriptors for each key point detected in the image scale-space. Descriptors proposed in [114] are based on local histograms of gradient directions. Matching these descriptors involves a threshold computed empirically from the images themselves. In order to make the method more robust, this threshold should be derived from statistical arguments. In this section, we intend to do so, by deriving the matching threshold following an *a contrario* approach. We shall propose a new SIFT descriptor for which an automatic matching strategy similar to the one presented in Chap. 5 can be applied. The new SIFT algorithm exhibits high matching efficiency and is able to detect several instances of the same object. And again, the clustering step proposed in Sect. 11.1 can be applied to further improve the results.

11.3.1 Normalization

Exactly the same key points as in [114] will be used in the proposed algorithm. We gave some details on the procedure to compute them in Sect. 10.1. Each key point comes with a position, but also with its scale and an orientation (which is one of the dominant gradient directions in the vicinity of the point). Hence it is characterized by an element $(x, s, \varphi) \in \mathbb{R}^2 \times \mathbb{R}_+ \times (-\pi, \pi)$. There are usually a few hundreds key points in a 512×512 image. Even though a very simple interpolation procedure attempts to refine the position of the key points, Lowe estimates that the position error is of the magnitude of the scale of the interest point, the error on the orientation is $\pm 15°$, and the scale is determined up to a $\sqrt{2}$ factor. It turns out that the accuracy is often much better than that, thus permitting a fair enough registration of the images. A pair of interest points (x, s, φ) and (y, t, ψ) in two images defines a unique similarity (four scalar parameters). This similarity writes

$$F(\xi) = zR\xi + b,$$

where $z = \frac{t}{s}$, R is the plane $\psi - \varphi$ rotation, and $b = y - zRx$. Let us assume that u is a gray level image and that v is obtained by applying the similarity F to u followed by a contrast change g (we recall that contrast changes are modeled by an increasing function $g : \mathbb{R} \to \mathbb{R}$). Thus, $v(x) = g(u(zRx + b)) = g(u(F(x)))$. An elementary calculation shows that

$$Dv(F(x)) = g'(u)zDu(F(x))R, \qquad (11.1)$$

meaning that the gradient of u has simply been rotated by the rotation R^{-1} and multiplied by a positive number. Therefore the direction of the gradient of v at point $F(x)$ is obtained by rotating the direction of the gradient of u at x.

11.3.2 Matching

Let us assume that u and v are two images, or pieces of images, of the same size belonging respectively to a set of images $\mathcal{Q}$ (for query) and $\mathcal{B}$ (for base). Let us denote by $N_\mathcal{Q}$ and $N_\mathcal{B}$ the cardinality of $\mathcal{Q}$ and $\mathcal{B}$. How to compare u and v and to come to the robust decision that they are similar? A similarity measure is often defined as a distance between descriptors. The comparison relies on the following fact: two images differing by a contrast change have the same gradient direction. On the contrary, if u and v are not related, then so are the directions of their gradient. In this case, it is sound to assume *a contrario* that the difference of these directions (in absolute value) is a uniform random variable in $(0, \pi)$, and independent of the values taken at remote enough points.

The *a contrario* approach consists in deciding that two images are actually similar when their observed similarity could not occur just by chance. More precisely, we have to check if the gradients of u and v are much more often aligned than the *a contrario* model can allow. For any point x, let us denote by $D(x)$ the difference of the directions of $Du(x)$ and $Dv(x)$. It is a number in the interval $(0, \pi)$, defined if both $Du(x)$ and $Dv(x)$ are nonzero. In order to avoid quantization effects on the gradient direction, only points where the gradient norm is larger than $\tau > 0$ can be considered. In practice, $\tau = 5$.

Let $x_1, ..., x_M$, be M points in the image domain of u. The way they are chosen will be the object of a further careful analysis. Let us consider the following *a contrario* hypothesis.

$\mathcal{H}_0$: the M values $(D(x_i))_{1 \leqslant i \leqslant M}$ are i.i.d., uniform in $(0, \pi)$.

Again, this hypothesis is clearly false if the images are similar. Our purpose is precisely to adequately reject this hypothesis. Let $\alpha \in (0, \pi)$, and $q_\alpha = \frac{\alpha}{\pi}$. The probability, under $\mathcal{H}_0$, that at least k among the M values $\{D(x_1), \ldots D(x_M)\}$ are less than α is given by the tail of the binomial law

$$B(M, k, q_\alpha) = \sum_{j=k}^{M} \binom{M}{j} q_\alpha^j (1 - q_\alpha)^{M-j}. \tag{11.2}$$

Otherwise said, $B(M, k, q_\alpha)$ is the probability that the directions of Du and Dv coincide at (at least) k points out of M, *by chance*. If for two images u and v, we indeed observe k such points and if $B(M, k, q_\alpha)$ happens to be very small, then chance is certainly not a good explanation. Let us note, however, that if the image sets $\mathcal{Q}$ and $\mathcal{B}$ are very large, then such an observation may indeed be casual. The fact that an observation should be considered as surprising (or not) depends on the size of the database. This leads us to the following definition, which follows the Desolneux et al. method [50]. For a detailed account, we refer to the book [54].

Definition 19. Let $0 \leqslant \alpha_1 \leqslant \ldots \leqslant \alpha_L \leqslant \pi$ be L values in $[0, \pi]$. For any $(u, v) \in \mathcal{Q} \times \mathcal{B}$, we call number of false alarms of (u, v) the quantity

$$\text{NFA}(u, v) = N_\mathcal{Q} \cdot N_\mathcal{B} \cdot L \cdot \min_{1 \leqslant i \leqslant L} B(M, k_i, q_{\alpha_i}), \tag{11.3}$$

where k_i is the cardinality of

$$\{j, \, 1 \leqslant j \leqslant M, D(x_j) \leqslant \alpha_i\}.$$

We say that (u, v) is ε-meaningful, or that u and v are ε-similar if $\text{NFA}(u, v) \leqslant \varepsilon$.

Since numbers of false alarms can be very small, the logarithmic scale is more intuitive and we call meaningfulness of (u, v) the value $\mathcal{M}(u, v) = -\log_{10}(\text{NFA}(u, v))$.

The interpretation of this definition will be made clear after stating the following proposition. We put its proof for a sake of completeness, but it it just a variant of the other meaningfulness propositions in the present book, in particular Props. 8 and 10.

Proposition 12. *For two image sets $\mathcal{Q}$ and $\mathcal{B}$ such that $\mathcal{H}_0$ holds, the expected number of ε-meaningful pairs is less than or equal to ε.*

Proof. For all i, let us denote by K_i the random number of points among the x_j such that $D(x_j)$ is less than α_i. For any v, (u, v) is ε-meaningful, if there is at least $1 \leqslant i \leqslant L$ such that $N_\mathcal{Q} \cdot N_\mathcal{B} \cdot L \cdot B(M, K_i, q_{\alpha_i}) < \varepsilon$. Let us denote by $E(u, v, i)$ this event. Its probability $P_{\mathcal{H}_0}(E(u, v, i))$ satisfies

$$P_{\mathcal{H}_0}(E(u, v, i)) \leqslant \frac{\varepsilon}{L \cdot N_\mathcal{Q} N_\mathcal{B}}.$$

Indeed, for any real random variable X with survival function $H(x) = \Pr(X > x)$, it is a classical fact that $\Pr(H(X) < x) \leqslant x$. By applying this result to K_i, we get the upper bound on $P_{\mathcal{H}_0}(E(u, v, i))$. The event $E(u, v)$ defined by "(u, v) is ε-meaningful" is $E(u, v) = \cup_{1 \leqslant i \leqslant L} E(u, v, i)$. Let us denote by $\mathbb{E}_{\mathcal{H}_0}$ the mathematical expectation under $\mathcal{H}_0$. Then

$$\mathbb{E}_{\mathcal{H}_0}\left(\sum_{u\in\mathcal{Q},\, v\in\mathcal{B}} \mathbf{1}_{E(u,v)}\right) = \sum_{u\in\mathcal{Q},\, v\in\mathcal{B}} \mathbb{E}_{\mathcal{H}_0}\left(\mathbf{1}_{E(u,v)}\right)$$

$$\leqslant \sum_{\substack{u\in\mathcal{Q},\, v\in\mathcal{B}\\ 1\leqslant i\leqslant L}} P_{\mathcal{H}_0}\left(E(u,v,i)\right)$$

$$\leqslant \sum_{\substack{u\in\mathcal{Q},\, v\in\mathcal{B}\\ 1\leqslant i\leqslant L}} \frac{\varepsilon}{LN_{\mathcal{Q}}N_{\mathcal{B}}} = \varepsilon. \qquad \square$$

Remark that if $\mathcal{Q}$ and $\mathcal{B}$ are white noise images, then $\mathcal{H}_0$ trivially holds. For two such bases, any detection is a false alarm. Indeed, it is *a priori* known that the images are unrelated, which does not mean that they have nothing in common. The number of false alarm quantifies what has to be accepted as a casual similarity. Hence, by setting $\varepsilon = 1$, one (false) detection may be observed on average in databases of noise images. Obviously, this still holds if only one of the images u and v is made of noise. It actually turns out that $\mathcal{H}_0$ is reasonable if the sample points $x_1, ..., x_M$ are carefully chosen, as discussed in Sect. 11.3.3.

Thus, Def. 19 together with Prop. 12 mean that there are, on average, less than ε pairs of images (u, v) in $\mathcal{Q} \times \mathcal{B}$ that match by chance, that is to say, when $\mathcal{H}_0$ holds. Under this hypothesis, any detection must be considered as a false alarm (hence the denomination of NFA). Thus, it is chosen to eliminate any observation having a frequency of the order of ε in the *a contrario model*.

The values q_α are simply quantization steps and are known *a priori*. Hence, it is possible to tabulate the values of the binomial law once and for all, and to rapidly compute the number of false alarms. It is possible to figure out the behavior of the NFA with respect to the parameters, thanks to the following asymptotic expansion, first proved by Hoeffding (for more details, see the original article [84] and the textbook [54]).

Proposition 13. *Let $H(r, p) = r\ln\frac{r}{p} + (1-r)\ln\frac{1-r}{1-p}$, be the relative entropy of two Bernoulli laws with parameters r and p. Then, for $k \geqslant Mp$,*

$$B(M, k, p) \leqslant \exp\left(-M \cdot H\left(\frac{k}{M}, p\right)\right). \tag{11.4}$$

This inequality leads to the following sufficient condition of meaningfulness.

Corollary 2. *If*

$$\max_{\substack{1\leqslant i\leqslant L\\ k_i\geqslant Mq_{\alpha_i}}} H\left(\frac{k_i}{M}, q_{\alpha_i}\right) > \frac{1}{M}\ln\frac{LN_{\mathcal{Q}}N_{\mathcal{B}}}{\varepsilon}, \tag{11.5}$$

the pair (u, v) is ε-meaningful.

In this corollary, it appears clearly that the value of k such that (u, v) is ε-meaningful only depends on the logarithm of L, $N_{\mathcal{Q}}$, $N_{\mathcal{B}}$ and ε. In practice, we choose L about 10 which is compatible with our perceptual accuracy of directions. We also take $\varepsilon = 1$ since it means that we have on average less than 1 false detection. But, as we shall see, really similar images have much smaller NFA and the choice of ε is not really important. Thus, in all experiments, we always set $\varepsilon = 1$, and we can therefore claim that the decision threshold is automatically derived. The asymptotic estimate given by Hoeffding's inequality [84] also shows the conditions to obtain a small number of false alarms. If M is fixed, then the NFA is a decreasing function of the proportion of coincidental direction $\frac{k}{M}$. If the proportion is assumed to be fixed, then the number of false alarms is exponentially (thus very fast) decreasing with respect to the number of samples M.

11.3.3 Choosing Sample Points

The computation of the number of false alarms is made under the assumption $\mathcal{H}_0$. Thus, we assume u and v to be independent randomly chosen images. But we also assume that the fact that $Du(x)$ and $Dv(x)$ are collinear is independent from the fact that $Du(y)$ and $Dv(y)$ are collinear at some other pixel y.

We must make this assumption realistic. There are two reasons for being careful. The first one is that if x and y are too close to each other, then $Du(x)$ and $Du(y)$ can be correlated. Thus, we must take pixels at a critical minimal distance to ensure independence of their gradients. Since the gradient is computed by a 2×2 finite difference scheme, sample points must be at least two pixels afar, in the original images. This has to be corrected by the scaling introduced by the normalization. If for instance, u has to be zoomed in by a factor 4 before comparison with v, then the minimal distance between two samples in the resulting image is $2 \times 4 = 8$.

The second issue for ensuring independence of observations is what we shall call the *alignment problem*. Images contain shapes, whose boundary are often piecewise smooth curves, that can locally be approximated by straight lines. The orientations of the gradient at two points on an alignment are the same, and cannot be assumed independent, even though the points may be far from each other. Hence sample points must be chosen sparse enough to minimize the probability that they fall on an edge which is common to both u and v. This obviously puts strong limits on the number of samples that may be drawn. In [32], calculations showed that the number of samples that are necessary to attain very low numbers of false alarms (yielding detection) is about 100, with reasonable noise conditions. When globally comparing two images, drawing about 100 samples yields very good results with a number of false detections which is conform to the prediction.

When dealing with image patches, the results may be less satisfactory. Indeed, patches result from a normalization procedure. Two parts of images basically containing an edge lead to two normalized patches that are very similar, and aligned with the edge direction. Moreover, if points are uniformly sampled over the sets of pixels with a large enough gradient, the gradient direction difference will be small with a high probability, since all the samples are very likely to belong to the (single) edge. Hence, the number of false alarms will be small. The patches are very similar indeed, and the detection is not a false alarm to this respect. However, it is not very informative. This problem is analog to the aperture problem in the estimation of optical flow: edges are locally indistinguishable.

This implies that the comparison of gradient directions is sensible only if the images are complex enough. An easy way to check this is to impose to draw samples in u and v such that the gradient direction in u (for instance) is close to uniform over the sample points. This way, samples are constrained not to lie on the same alignment and to restore some of the complexity of the images. These considerations lead us to the sampling algorithm described in the next section.

11.4 The Detection Algorithm

For two images I_1 and I_2 compute their interest points $(x_i, s_i, \varphi_i), (y_j, t_j, \psi_j)$. With no loss of generality, assume that $t_j \geqslant s_i$ (else reverse the role of I_1 and I_2). Let F be the similarity $F(x) = zRx + b$ mapping (x_i, s_i, φ_i) on (y_j, t_j, ψ_j), where $z > 0$, R is a plane rotation, and $b \in \mathbb{R}^2$, are uniquely determined. Define two normalized images u and v. The image v is obtained by cropping I_2 in a patch $\mathcal{P}$ around y_j. The image u is simply $I_1 \circ F^{-1}$, also restricted to $\mathcal{P}$.

Let us quantize the direction of Dv in $(0, 2\pi)$ on $2N$ values. Sample points are then chosen by the following recursion. Let us consider $x_1 \in \mathbb{R}^2$ such that $|Du(x_1)| > \tau$ and $|Dv(x_1)| > \tau$. Assume that n points have been sampled. Then, a $n + 1$th point x_{n+1} is chosen such that

- $|Du(x_{n+1})| > \tau$ and $|Dv(x_{n+1})| > \tau$.
- For each k, $0 \leqslant k < 2N$, the number of points of $\{x_1, ..., x_{n+1}\}$ such that the direction of Dv belongs to the interval $\left(\frac{k\pi}{N}, \frac{(k+1)\pi}{N}\right)$ is less than $1 + \frac{n}{2N}$.

This simply means that the repartition of the sample points is required to stay essentially uniform. This is not always possible. If the histogram of directions in v is unimodal, then no long sequence will fit the condition. As a consequence, if K points are tested, then they may lead to a set of sample points containing much less than K points. Let M be this number of points. The number of false alarms between u and v is then computed by using Def. 19. Hence, the final number of samples depends on the complexity of v. If v is essentially an edge, the sample sequence will be short and the number of false alarms will be large (since computed on very few points). On the contrary, for complex images (for instance pure texture), sample sequences will be long, thus leading to very small NFAs.

Let us remark that this peculiar sampling introduces a slight asymmetry in the algorithm between u and v and that the algorithm would be strictly contrast invariant only if $\tau = 0$. In practice, letting $\tau = 5$ removes gray level quantization effects which can entail false detections [49].

Numerically, experiments were led with $K = 1000$, and M is restricted to be less than 300. This means that points are drawn according to the two conditions above. The loop ends either when the total number of drawn points reaches $K = 1000$

or when the number of admissible points reaches 300. In practice the number of admissible points ranges from 50 to 300. When the patches present the alignment problem, M is obviously small. In this case, numbers of false alarms are large, as can be seen on the asymptotic development (11.4).

11.4.1 Experiments: Securing SIFT Detections

In the experiments, $\mathcal{P}$ is a square patch whose size is proportional to the key point scale (a factor 25 is used).

In all the experiments, we observed the following facts:

- Right matches can be very meaningful $(-\log_{10}(\text{NFA}) \simeq 30)$;
- on the contrary the wrong matches meaningfulness is close to 0. When it is not the case, there is actually a strong geometrical similarity between the compared patches.

Fig. 11.8 Multiple matches experiment

In the experiment of Fig. 11.8 and 11.9, there are several partial occurrences of a logo. The usual procedure for matching SIFT descriptors (nearest neighbor, then comparison with second nearest neighbor) is inefficient in this case. In this experiment, 72 matches with an NFA less than 1 are detected. The best match has a meaningfulness (*i.e.* $-\log_{10}(\text{NFA})$) equal to 25.4. Half the matches have NFA less than 10^{-3}, and are of course correct.

That there can be meaningful but wrong matches is not only inevitable but semantically sound. Fig. 11.12 shows such an example. Clearly the two matched SIFT patches do not correspond to the same motorcycle. All the same both present a wheel with the same orientation, taken from the same perspective and under the same light.

Fig. 11.9 Matching groups in the images of Fig. 11.8. Each circle represents a SIFT key point which got a match. The radius is twice the scale of the key point. The grouping procedure is the same as in in Chap. 8

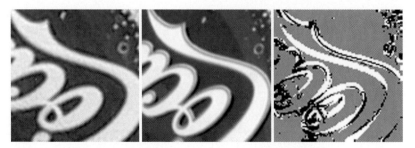

Fig. 11.10 Best SIFT match between the images of Fig. 11.8. Its meaningfulness is 25.4. On the left, the two registered patches. On the right figure, points where the gradient is not larger than 5 in both patches are displayed in gray. When the gradient is large in both images and when directions of the gradient coincide up to $40°$, the pixel is plotted in white. It is plotted in black otherwise. Even though the images are different, the registration provided by the key points is accurate enough and yields a very meaningful match

Fig. 11.11 A false and meaningful SIFT match between the two patches displayed on the left. The meaningfulness is 2.4. As can be seen, there are many white dots. Moreover, the orientation of the gradient at these locations is clearly not unimodal and the detection is not due to the presence of a single alignment

Under such circumstances, it is not only acceptable, but even desirable to make the detection. This is clearly illustrated in Fig. 11.8 which displays many physically different but identical Coca Cola cans. Fig. 11.10 shows an excellent match between two of them, taken from the same image. As we commented in Sect. 10.3.2.4 the SIFT original threshold procedure discards such matches. However, they are obviously of high interest. The detection and grouping of similar shapes in the same image is actually a fundamental gestalt [95]. However, some meaningful (but usually not very meaningful) casual matches can occur, which do not correspond to any recurrence of the same shape. Fig. 11.11 shows such a match between two SIFT patches, with 2.4 meaningfulness. Clearly the gradient orientations are very similar in both patches at many pixels, and the patches are complex enough to make such a coincidence an unlikely event. The overall explanation of such coincidence is given by the Gestalt Theory, which points out the recurrence of standard shapes in most images. Such shapes are called *gestalts* and include among others convex curves and bars with constant width [95]. The fact that two different patches show a similar arrangement of frequent shapes is therefore not unlikely. But this experiment proves that the *a contrario* model for patches similarity developed in this chapter is still a bit too primitive.

Fig. 11.12 Stroboscopic effect due to similarity within a class of objects. The second image has been obtained by a real defocus of the camera. On the bottom left, two patches that are very alike. The meaningfulness of their SIFT match is 2.2. The gradient directions difference is less than 40° at many locations (dots in white in the bottom patch). These patches do not come from exactly the same object, but retrieving them cannot be considered as a false alarm. Original images courtesy of the LEAR Team, INRIA

The next experiment shows matching and grouping results between two different views of the church of Valbonne (Fig. 11.13). Because of parallax, two different groups are detected, which is correct.

Fig. 11.13 Two different views of Valbonne church. There are two groups, corresponding to two different parts of the scene with different depths. In the first group, the largest meaningfulness is 30.8, in the second group, 2.7 because the resolution is lower in this part

11.5 Bibliographic Notes

Following [109, 139] the fact that some complex enough element recurs in two different images or even in the same can be taken as a basic definition of shape. Shapes simply are parts of an image which can be recognized in another one. There have been of course several attempts to measure the certainty of detected shape similarities. Schmid et al. [158, 157] use statistics of the distance between descriptors to recognize parts of objects of the same type, in a semi-supervised way. The grouping of SIFT matches for attaining certainty was pointed out in Lowe [113] and more recently in Cao et al. [33]. The grouping phase in [114, 157] is used *a posteriori*, both to eliminate possible casual matches, and to reinforce the detection of right matches. The novelty in [33] is the accurate computation of the number of false alarms assigned to a group. An alternative to the method described in Sect. 11.3 and 11.4 for meaningful SIFT matching has been recently proposed in [150].

Appendix A
Keynotes

A.1 Cluster Analysis Reader's Digest

Clustering methods have been and are still the object of applied and theoretical research in many different fields such as statistical pattern recognition, data mining, image processing, biomedical sciences, etc. It is not the aim of this section to present a complete overview of clustering techniques, but rather to provide enough information to justify why a particular technique may be preferred (there is no universal best clustering algorithm, and choices and compromises have to be made). A good review of clustering techniques by Jain *et al.*, from a statistical pattern recognition viewpoint, can be found in [93]. The main concepts can also be found in Duda and Hart [60], Hastie *et al.* [82] and Kaufman and Rousseeuw [96] textbooks.

Most of the clustering algorithms are either partitional, or hierarchical methods. While partitional methods produce a single partition, hierarchical methods produce a nested series of partitions. In this sense, they provide a totally different data description and should not be considered as two competing techniques. However, as shall be seen, because of their different nature, the corresponding strategies for cluster validity assessment may be quite different.

A.1.1 Partitional Clustering Methods

Let us denote by $\mathcal{T} = \{T_k, \ k \in \{1, \dots, M\}\}$ the data set where each pattern T_k is a D-dimensional feature vector, and by $d_T : \mathcal{T} \times \mathcal{T} \to \mathbb{R}^+$ the dissimilarity measure. Assuming for the moment that the partition size c is given, the goal of a partitional clustering algorithm is to identify the partition $\mathcal{P}(\mathcal{T}) = \{\mathcal{T}_1, \dots, \mathcal{T}_c\}$ on $\mathcal{T}$ that optimizes a criterion function. Parametric methods as mixture decomposition will not be addressed here since there is no a *a priori* knowledge on the underlying probability distribution. (In these methods, the data set is assumed to be drawn from a mixture of c underlying parametric distributions, and the goal is to determine the involved parameters. The standard algorithm is the Expectation-Maximization algorithm [46].) Hence, since there are approximately $c^M/c!$ ways of partitioning a set of M elements into c subsets (a Stirling number of the second kind), optimizing the criterion function by exhaustive search is intractable and iterative optimization procedures are needed.

The simplest and most widely used family of criteria function is the one of related minimum variance criteria [60, 96]. The energy to be minimized here is

$$E = \frac{1}{2} \sum_{m=1}^{c} n_m \langle d_m \rangle,$$

where n_m is the number of points in the m-th cluster, and

$$\langle d_m \rangle = \frac{1}{n_m^2} \sum_{T_i \in \mathcal{T}_m} \sum_{T_j \in \mathcal{T}_m} d_T(T_i, T_j)$$

is the average dissimilarity measure between points in the m-th cluster. If $\mathcal{T}$ was a subset of a vector space, and d_T was the squared Euclidean distance, the resulting criteria would be the sum of variances of each cluster,

$$\sum_{m=1}^{c} \sum_{T \in \mathcal{T}_m} \| T - \langle T_m \rangle \|_2^2, \quad \text{where } \langle T_m \rangle = \frac{1}{n_m} \sum_{T \in \mathcal{T}_m} T.$$

Strictly speaking, this criterion only makes sense when clusters are isotropic, multivariate normally distributed. Moreover, the solution is not invariant to linear transformations of the data. Many variations on this method exists, taking any Minkowski metric or the squared Mahalanobis distance instead of the squared Euclidean distance [93]. Notice however that all of these methods are based on the notions of medoid or centroid (barycenter) of a set of points and this does not make sense unless patterns live in a vector space.

Related minimum variance criteria suffer from the problem that partitions that split large clusters may be favored over ones that maintain the integrity of natural clusters [60]. When natural clusters have very different number of points, the partition minimizing this criteria may not reveal the intrinsic structure of the data. Another weakness of these methods is the lack of ability to extract a very dense cluster embedded in the center of a diffuse cluster. Besides, the partition solution has to be found by iterative optimization procedures. These iterative procedures are to be initialized by a reasonable initial partition and solution can be trapped in local minima [93].

Other popular criterion functions, also defined only when patterns live in Euclidean (or Hermitian) spaces, and closely related to the previous ones, can be derived based on the within cluster scatter matrix $W\left(\mathcal{P}(\mathcal{T})\right)$, and the between cluster scatter matrix $B\left(\mathcal{P}(\mathcal{T})\right)$ [60],

$$W\left(\mathcal{P}(\mathcal{T})\right) = \sum_{m=1}^{c} \sum_{T \in \mathcal{T}_m} \left(T - \langle T_m \rangle\right) \cdot \left(T - \langle T_m \rangle\right)^{\mathrm{T}},$$

$$B\left(\mathcal{P}(\mathcal{T})\right) = \sum_{m=1}^{c} n_m \left(\langle T_m \rangle - \langle T \rangle\right) \cdot \left(\langle T_m \rangle - \langle T \rangle\right)^{\mathrm{T}},$$

$$S = \sum_{m=1}^{c} \left(T - \langle T \rangle\right) \cdot \left(T - \langle T \rangle\right)^{\mathrm{T}} = W\left(\mathcal{P}(\mathcal{T})\right) + B\left(\mathcal{P}(\mathcal{T})\right),$$

where $\langle T \rangle$ is the barycenter of all patterns in the data set, and S is the total scatter matrix, which is a constant given the data, independent on the partition. One can define optimal partitions as minimizers of $\mathrm{tr}\left[W\left(\mathcal{P}(\mathcal{T})\right)\right]$ (or equivalently maximizers of $\mathrm{tr}\left[B\left(\mathcal{P}(\mathcal{T})\right)\right]$). This turns out to be a minimum variance criterion. Another possibility is to minimize $\det\left[W\left(\mathcal{P}(\mathcal{T})\right)\right]$ whose solution is invariant to linear transformations of the data. In any case, combinatorial optimization is intractable and one has to consider iterative procedures.

A.1.2 Iterative Methods for Partitional Clustering

Most partitional methods are based on the definition of c elements from a pattern space called centrotypes, each defined to be a representative object of one of the clusters. The criterion function to be minimized is usually the sum of the average dissimilarities between each centrotype and all of the other patterns of the same cluster. Typically, iterative methods begin by initialising a set of c centrotypes. Each pattern is then assigned to the cluster corresponding to its closest centrotype (for the considered dissimilarity measure), and centrotypes are re-computed in order to mimimize the criterion function. The iteration ends when centrotypes do not change. The computational efficiency of this approach depends on how easily centrotypes can be computed. The c-means algorithm [115] (also referred in the literature as k-means) runs typically in $O(M)$ [25]. Indeed in this algorithm the dissimilarity measure is the squared Euclidean distance and centrotypes are the clusters' barycenters, which can be easily computed using an update equation. A similar algorithm can be obtained by using the ℓ_1-norm as dissimilarity measure. The centrotypes for this measure (which is more robust to outliers than the squared Euclidean distance) are the cluster medians.

When the dissimilarity measure does not lead to a closed form representation for the centrotypes, a method known as k-medoid which allows clustering with respect to any specified dissimilarity measure can be used [96]. In this method, centrotypes (the so-called medoids) are restricted to be patterns from the data set, and as before patterns are assigned to the cluster corresponding to its closest centrotype. The goal is then to select, among all M patterns the c centrotypes which minimize the sum of the average dissimilarities between each centrotype and all of the other patterns of the same cluster. A widely used implementation for the k-medoid method is the

Partitioning Around Medoids algorithm (PAM), by Kaufman and Rousseeuw [96]. PAM consists of two phases. In the first one, a method for selecting the initial set of c centrotypes or medoids is applied. The second phase is an iterative procedure where in each iteration the set of centrotypes is updated by analyzing all possible pairs of patterns such that one pattern is a centrotype and the other is not, and by swapping the pair which most reduces the value of the criterion function. The cost of a single iteration is $O(c(M - c)^2)$.

A.1.3 Hierarchical Clustering Methods

While partitional clustering algorithms construct a single partition with c clusters (a flat description), hierarchical methods deliver a recursive structure. Since they represent data in different ways, partitional and hierarchical methods do not really compete with one another. Indeed, when data is to be described in terms of classes, subclasses, sub-subclasses (e.g. a biological taxonomy), flat representations do not make sense and hierarchical methods are needed. There are, of course, many applications in which data is not inherently hierarchical, and one has to make a choice among clustering methods from both types. Hierarchical methods are more versatile than partitional methods and can deal with many differently shaped clusters, but generally they are more time consuming.

Depending on the direction they build the hierarchy, these clustering methods can be agglomerative (bottom-up) or divisive (top-down). The former, which are usually computationally simpler, start with each single point as a cluster, and iteratively merge the closest pair of clusters in the sense of a chosen dissimilarity measure. The generic algorithm is as follows [93].

1. Initialization: compute the proximity matrix (the matrix containing the dissimilarity between each pair of patterns).
2. Find the most similar pair of clusters using the proximity matrix. Merge these two clusters.
3. Update the proximity matrix according to this merging.
4. Repeat steps 2 and 3 until all patterns are in one cluster.

At each iteration step, two clusters are merged. The procedure builds up a tree or dendrogram, where leaves are the M elements of $\mathcal{T}$ (step 1). At level l, this tree has $M - l$ nodes, each node being a cluster. At level $l + 1$, the closest clusters from level l are merged (step 2). By closest, we mean the pair $\mathcal{T}_i$ and $\mathcal{T}_j$ minimizing a given distance or proximity measure $\delta(\mathcal{T}_i, \mathcal{T}_j)$ between clusters. Different strategies for updating the proximity matrix lead to different hierarchical clustering methods. (Moreover, since all of these algorithms are merging methods, they admit a variational formulation and can be solved as an energy minimization problem; see [137], Chap. 3.) Lance and Williams [103] define a class of methods by specifying a generalized recurrence formula for updating the proximity matrix:

$$\delta(\mathcal{T}_i \cup \mathcal{T}_j, \mathcal{T}_k) = \alpha_i\, \delta(\mathcal{T}_i, \mathcal{T}_k) + \alpha_j\, \delta(\mathcal{T}_j, \mathcal{T}_k) + \beta\, \delta(\mathcal{T}_i, \mathcal{T}_j) + \gamma\, |\delta(\mathcal{T}_i, \mathcal{T}_k) - \delta(\mathcal{T}_j, \mathcal{T}_k)|,$$

where parameter values α_i, α_j, β and γ characterize the particular clustering method. Below we describe the most popular ones.

- Choosing $\alpha_i = \alpha_j = 1/2$, $\beta = 0$ and $\gamma = -1/2$, leads to the following distance between clusters:

$$\delta_{min}(\mathcal{T}_p, \mathcal{T}_q) = \min_{T_i \in \mathcal{T}_p, T_j \in \mathcal{T}_q} d_T(T_i, T_j).$$

  The corresponding algorithm is known as *single-linkage algorithm* [93, 60]. Here the nearest-neighbor points determine the nearest subsets. If elements in $\mathcal{T}$ are viewed as nodes of a graph, merging $\mathcal{T}_p$ and $\mathcal{T}_q$ corresponds to adding an edge between the nearest points in $\mathcal{T}_p$ and $\mathcal{T}_q$. This procedure generates a tree, and if one lets the procedure evolve up to having a single cluster containing all points, the result is a *minimal spanning tree*.
- Taking $\alpha_i = \alpha_j = \gamma = 1/2$, $\beta = 0$, yields

$$\delta_{max}(\mathcal{T}_p, \mathcal{T}_q) = \max_{T_i \in \mathcal{T}_p, T_j \in \mathcal{T}_q} d_T(T_i, T_j).$$

  The resulting algorithm is called *complete-linkage algorithm* [93, 60]. Here distance between two clusters is given by the farthest pair of points in the two clusters. This procedure produces a graph in which edges connect all of the nodes in a cluster. When the nearest clusters are merged, edges between every pair of nodes in the two clusters are added. If the diameter of a partition is defined as the largest diameter for clusters in the partition, then each iteration of the complete-linkage algorithm increases the diameter of the partition as little as possible.
- Taking $\alpha_i = n_i/(n_i + n_j)$, $\alpha_j = n_j/(n_i + n_j)$, and $\beta = \gamma = 0$, leads to a group averaging method, where

$$\delta_{avg}(\mathcal{T}_p, \mathcal{T}_q) = \frac{1}{n_p n_q} \sum_{T_i \in \mathcal{T}_p} \sum_{T_j \in \mathcal{T}_q} d_T(T_i, T_j).$$

- Some clustering methods based on barycenters, such as Ward's minimum variance method [176], can also be represented in terms of Lance and Williams formula. For Ward's method, $\alpha_i = (n_i + n_k)/(n_i + n_j + n_k)$, $\alpha_j = (n_j + n_k)/(n_i + n_j + n_k)$, $\beta = -n_k/(n_i + n_j + n_k)$, $\gamma = 0$, and the corresponding cluster proximity measure is

$$\delta_{ward}(\mathcal{T}_p, \mathcal{T}_q) = \frac{n_p n_q}{n_p + n_q} \|\langle T_p \rangle - \langle T_q \rangle\|_2^2,$$

where $\langle T_p \rangle$ and $\langle T_q \rangle$ denote the barycenters of $\mathcal{T}_p$ and $\mathcal{T}_q$ respectively.

Time and memory complexity of the algorithms given by the Lance and Williams formula are studied in [44]. Overall, the time required for hierarchical clustering is $O(M^2 \log M)$, and the memory complexity is $O(M^2)$.

In practice, if clusters are compact and well separated, all methods yield the same results. However, when this is not the case, the resulting partitions may be quite different. Depending on the cluster proximity measure, different methods of clustering can be more or less successful with different types of clusters. Single-linkage algorithms suffer from the chaining effect: A single corrupted point somewhere in between two compact clusters can lead to an unwanted merging between them [93, 60]. However, this property is very useful if one wants to detect elongated clusters.

The complete-linkage algorithm tends to produce compact clusters with small diameters. However, patterns assigned to a cluster can be much closer to patterns in other clusters [82, 60].

The single-linkage and the complete-linkage algorithms are both sensitive to outliers since they rely on extremal measures. One way to reduce the influence of outliers is using δ_{avg} as cluster proximity measure though the improvement is often not good enough. Besides, average methods have another drawback compared to single or complete linkage methods: they are not invariant under monotone transformations on the dissimilarity measure d_T (invariance of the former ones is a consequence of being based on extremal values) [82].

To end this section, let us make a few general remarks. In Sect. A.1.1, one of the main assumptions is that the number of clusters c was given, for partitional clustering algorithms. Then, the goal was to find the c-partition on the data optimizing a global criterion (in practice iterative methods are used and the convergence to a global minimum is not ensured). Agglomerative hierarchical clustering methods perform well in making local decisions about cluster merging since they make use of the proximity matrix. As the hierarchy is built by means of local optimization, the level corresponding to a c-partition will not correspond in general to a global optimum (unless clusters are compact and well separated). For instance, Ward's method will not lead to the same c-partition as a c-means method, despite the fact that both attempt to minimize variance. In this sense, one would rather say that partitional methods are better than hierarchical methods. But how to be sure that there are exactly c groups of patterns in the data? Is the criterion function well adapted to the shape of clusters that are present in the data? From this viewpoint, hierarchical clustering may be more appealing than partitional ones. Another argument in favor of hierarchical clustering methods is their versatility and their ability to cope with differently shaped clusters. For instance, the single linkage algorithm can deal with non-isotropic, elongated or concentric clusters while partitional methods like c-means can only deal with isotropic clusters. Since their outputs are nested series of partitions, ranging from M clusters to one single cluster, one can imagine methods to determine the number of clusters as stopping rules in the merging process. If stopping rules are correctly designed, hierarchical methods would also be able to detect clusters having different densities or different number of points.

A.1.4 Cluster Validity Analysis and Stopping Rules

The great variety of clustering methods that have been proposed in the recent past has been followed by an increasing interest in clustering validation methods. In [73], a comprehensive study of these techniques is presented.

Cluster validity analysis deals with assessing the validity of classifications obtained from the application of clustering procedures. There are different validation approaches [58, 73] depending on the amount of prior information on the data. This section deals with *internal validation tests*, which consist in determining if the structure is intrinsically adapted to the data. In other words, internal tests are derived from some *internal criteria* measuring the suitability of the clustering structure for the original data set with no other information than the data themselves.

Classical issues in cluster validity analysis are the assessment of individual cluster validity and the assessment of a whole partition. (In some applications validity of a dendrogram also needs to be assessed. This problem is not addressed here.) These two issues are briefly summarized next.

A.1.4.1 Partition Validity Assessment

A relevant question to address in order to assess the validity of a partition, is deriving the number of clusters [58], denoted by c. Notice that by solving this problem, it cannot be ensured that the c clusters are valid clusters. The most common approach to decide how many clusters are best consists in finding partitions for $c = 1, \ldots, c_{max}$ and optimizing a measure $G(c)$ of partition adequacy, which is usually based on the within-cluster and between-cluster variability. When applied to hierarchical clustering methods these cluster validity assessment techniques are known as *global stopping rules* because the choice of c can be seen as stopping the merging process (in the agglomerative case) at a certain level of the dendrogram.

When dealing with hierarchical classifications, another approach to determine the most appropriate number of clusters are *local stopping rules*. In the agglomerative case, these rules are *merging criteria* to decide whether two clusters should be merged. Usually, the merging process is continued until it is decided, for the first time, that two clusters should not be aggregated.

Milligan and Cooper [125], and Dubes [58], present comparative studies of some stopping rules. Milligan and Cooper's paper provides a particularly comprehensive Monte-Carlo evaluation of these rules, by comparing thirty local and global stopping rules. In their simulation experiment, only strongly clustered data sets (internally cohesive and well separated clusters) were considered. Hence, since clustering this kind of data should not be a challenging problem, techniques that do not perform well on it are also expected to be inefficient when dealing with any data set. The main conclusion of this experiment is that only five or maybe six of the compared rules perform quite well on strongly clustered data. One can also observe that the majority of the stopping rules described in the study are based on heuristics and lack of theoretical foundation. Those derived from rigorous statistical techniques, assume in general hypotheses on the data which are unrealistic in most real applications (*e.g.* multivariate normal distribution for the patterns). In order to briefly illustrate the considered stopping rules, it is worth describing Calinski and Harabasz's index [26] and Duda and Hart's rule [60], since these methods provided the best results.

- Calinski and Harabasz propose a *global stopping rule* for assessing partitions, by choosing the partition size c that maximizes the index

$$G(c) = \frac{\frac{1}{c-1}\mathrm{tr}\left[B\left(\mathcal{P}(\mathcal{T})\right)\right]}{\frac{1}{M-c}\mathrm{tr}\left[W\left(\mathcal{P}(\mathcal{T})\right)\right]},$$

  where $B\left(\mathcal{P}(\mathcal{T})\right)$ and $W\left(\mathcal{P}(\mathcal{T})\right)$ are respectively the between- and within-cluster scatter matrices of a c-partition $\mathcal{P}$, defined in section A.1.1. The index $G(c)$ is the ratio between the total within-cluster sum of squared distances about the centroids, and the total between-cluster sum of squared distances. This index is only defined for sets of patterns living in an Euclidean space. Moreover, since the index is based on the sum of squares criterion, it has a tendency to partition the data into hyperspherical shaped clusters, having roughly equal numbers of patterns [73] (this is probably the main reason for its first position in Milligan and Cooper's ranking, since their data was strongly clustered, and clusters contained almost the same numbers of points and were pretty isotropic).
- Duda and Hart proposed the $Je(2)/Je(1)$ *local stopping rule* for deciding whether or not a cluster should be split into two subclusters. The rule consists in computing the ratio between the total within sum of squared distances about the centroids of the two clusters for the two-cluster solution ($Je(2)$), and the within sum of squared distances about the centroid when only one cluster is present ($Je(1)$). The method considers a null hypothesis, assuming that all patterns come from a normal distribution, whose mean and variances are empirically estimated over the whole data set. The null hypothesis of one single cluster is rejected if $Je(2)/Je(1)$ is smaller than a specified critical value, fixed by a significance level for the hypothesis testing. While considering a normal distribution as a null hypothesis and using the sum of squared distances may not be well adapted to real clustering problems (particularly when the number of patterns in the data set is not as large to be well represented by an asymptotic distribution), the proposed *a contrario* formulation is appealing from our point of view.

To finish the discussion on partition validity assessment we quote one of Bock's conclusions from his work on significance tests in cluster analysis [24], where a comparison between global and local methods is made.

> *Some care is needed when applying any test for clustering, bearing in mind that different types of clusters may be present simultaneously in the data, and that the number of clusters is, in some sense, dependent on the intended level of information compression. Thus, a global application of a cluster test to a large or high-dimensional data set will not be advisable in most cases. However, a local application (...) to a specific part of the data will often be useful for providing evidence for or against a prospective clustering tendency.*

A.1.4.2 Validity Assessment of Individual Clusters

The problem is now to decide, among the candidate clusters furnished by the clustering procedure, which ones correspond to natural clusters. But what does a natural cluster look like? As pointed out by Gordon [73], it may be difficult to specify a relevant definition of an ideal cluster for a particular data set. However, clusters must reveal structure in the data and can be detected as opposed to a complete absence of structure. Thus, in order to decide whether the candidate clusters are significant, they can be compared to some appropriate random distribution. This leads to a general methodology for cluster validity analysis based on the statistical approach of hypothesis testing [24, 72, 73]. Following Bock [24], this framework consists of these stages:

1. Design a null hypothesis $\mathcal{H}$ for the absence of class structure in the data (a *background model*, or *null model*), meaning that patterns are sampled from a homogeneous population. Then, heterogeneity or clustering structure are involved in the alternative hypothesis $\mathcal{A}$.
2. Define a test statistic, which will be used as a validity index to discriminate between $\mathcal{H}$ and $\mathcal{A}$.
3. If, for a given significance level (error probability) α, the test statistic of the observed data exceeds the corresponding critical value c_α, the null hypothesis $\mathcal{H}$ is rejected, in favor of $\mathcal{A}$.

This general framework can be adapted for assessing the validity of individual clusters. A general approach within this framework is the Monte-Carlo validation, which is described in [73]. Assume one wants to assess the validity of an observed cluster $\mathcal{T}_i$ having n patterns in a data set having M patterns. In the Monte-Carlo validation method, data sets of M patterns are simulated under the background model, and classified using the same clustering procedure that was used to classify the original data. The test statistic is computed for those clusters having n patterns, and the distribution of the test statistic is estimated. Then, using the value of the test statistic of $\mathcal{T}_i$, one can compute the significance level of rejecting $\mathcal{H}$. Two popular test statistics are the maximum F test and the U statistic (see Bock [24] and Gordon [73]).

Appropriate null models for data are the subject of the study presented in [72]. These models, which specify the distribution of patterns in the absence of structure in the data, can be of two types.

– *Standard (data-independent) null models.* Two well known standard null models are the *Poisson model* and the *Unimodal model* [24]. The main problem with the Poisson model is the choice of the region R within which patterns are uniformly distributed (standard choices for normalized data are the unit hypercube and the unit hypersphere). The Unimodal model assumes that the joint distribution of the variables describing the patterns is unimodal, but the choice of the distribution may not be easy.
– *Data-influenced null models.* Here the data is used to influence the specification of the null model. Examples of these null models are the Poisson model where R is chosen to be the convex hull of the data set, or the *Ellipsoidal model*, which is

a multivariate normal distribution, whose mean and covariance matrix are given by the data set.

In [72], Gordon concludes that the results of the tests depend considerably on the choice of the null model and that in general the results based on data-influenced null models are more relevant than those obtained using a standard null model.

A.2 Three classical methods for object detection based on spatial coherence

This section addresses some issues of the generalized Hough transform [14], whose variations are probably the most widely used techniques in object detection. Two frequently used techniques for robust transformation estimation will also be described: geometric hashing[102, 184] and the RANSAC algorithm [64].

A.2.1 The Generalized Hough Transform

In [14] Ballard proposed a generalization of the Hough transform [85] allowing the detection of arbitrary planar shapes undergoing similarity transformations. Most object detection and recognition systems using transformations clustering are based on the generalized Hough transform. The basic idea is to quantize the transformation space into D-dimensional cells. Each transformation point T_i is quantized and then votes for one of these cells. In practice, noise and image quantization induce localization errors in the extracted features and one has to take into account uncertainty in computing T_i. Thus, each pairing of model and image features defines a volume of possible transformations, so it should cast a vote into each cell intersecting this volume (see [75] for an error analysis when using line segments as features).

As with all techniques based on histograms in multidimensional spaces, the generalized Hough method is very sensitive to the choice of quantization precision (this remark also holds for Lamdan and Wolfson's Geometric Hashing [184, 102] described in Sect. A.2). Most of the time, the cell size is chosen by problem specific *ad hoc* arguments (see [113] for an example). However, in the general case, quantization effects may lead to several problems:

- Similar transformation points may vote for different cells. In order to reduce this problem, either votes are counted by adding the votes of neighboring cells (using a sliding window) in the case of no uncertainty in T_i, or, when uncertainty is considered, a vote is cast into each cell intersecting the uncertainty volume.
- In the plane similarity case, for instance, if one wants to do a fine discretization of the 4-D transformation space in order to perform accurate detection, the search space is too large for an exhaustive search. Coarse to fine techniques applied to transformation clustering, first introduced by Stockman [169], can deal with this complexity problem, but there is no reason why the most voted cells at the finer scale should correspond to the most voted ones at coarser scales.
- From the detection viewpoint, the size of the cells is also crucial. Indeed, if quantization is too fine, cells will not have enough votes and correct instances will be missed (false negatives). On the other hand, choosing a very coarse quantization increases the likelihood of large clusters occurring at random (false positives).

These remarks partially motivate our decision to use the clustering techniques described in Chap. 7, along with the validity assessment method proposed in the same chapter. Indeed, the proposed methodology does not suffer from quantization problems.

The generalized Hough transform is with geometric hashing [102, 182, 184], and the alignment method [89] one of the most popular voting schemes. Given two shapes, the geometric hashing method aims at determining if there is a transformed subset of the features from one shape that matches a subset of the features of the other one. The alignment method is a similar voting method. The generalized Hough transform method, instead of voting over all possible configurations of shapes, consists in voting over all possible transformations mapping a shape to another one. As for all techniques based on histograms in multidimensional spaces, these voting methods are very sensitive to the choice of quantization precision (too large bins may lead to false matches, and too small bins may produce misses). Besides, most of the time, the size of the hash table and the amount of parameters (the size of the bins in the voting stage, the threshold for the amount of votes in each bin, *etc.*) are crippling. The complexity of these voting schemes increases with the invariance degree; affine invariant shape retrieval in large databases is intractable. All these properties make the local features not suitable for shape retrieval in large databases.

A.2.2 Geometric Hashing

In order to illustrate the geometric hashing algorithm, we present the case of similarity or affine transformations.

A query shape S is searched in a set of shapes.

Preprocessing (off line). For each shape S_i' in the set of shapes:

1. Extract local invariant features from S_i'. Assume n such features are found.
2. For each local basis b_j (*e.g.* a pair of points for similarity transformations, three non-collinear points for affine transformations) of features:

   a. Compute the quantized coordinates (u, v) of all the remaining features, in the local basis.
   b. Use the couple (u, v) as an index in a hash table, and write the information (i, b_j) in the corresponding bin (i is the index that identifies S_i').

Recognition stage (on line). For the query shape S:

1. Extract local invariant features from S. Assume n such features are found.
2. Choose arbitrarily a local basis (two or three points, depending on the considered invariance).
3. Compute the quantized coordinates (u, v) of all the remaining features, in the local basis.
4. For each of these coordinates, go to the corresponding bin in the hash table, and cast a vote for each pair (i, b_j) inscribed in the bin.

5. Keep only the pairs (i, b_j) which received more than a certain number of votes. Each of these pairs stands for a potential match.
6. For each potential match, compute the best transformation (in the least squares sense) between all corresponding features, and check if the query features and the features from the corresponding shape are well aligned. If not, go to (2) and choose another basis.

For affine invariant shape recognition, time complexity for the preprocessing stage is $O(n^4)$ for each shape in the set of shapes. If the access time to the hash table is $O(1)$, time complexity for the recognition stage is between $O(m)$ (when the first query basis chosen at random corresponds to a model in the set of shapes) and $O(m^4)$ (when no basis from the query shape corresponds to a model in the set of shapes).

A.2.3 A RANSAC-based Approach

The RANdom SAmple Consensus (RANSAC) algorithm by Fischler and Bolles [64], is certainly one of the most popular robust estimators in computer vision. It has proved very successful in stereo vision tasks, such as estimating homographies and fundamental matrices [81]. The main reason for its success is its ability to deal with large proportions of outliers. Roughly speaking, in its general form, the RANSAC procedure to fit a model consists in randomly selecting a minimal subset of the data (*i.e.* a subset allowing to instantiate the model), then computing the number of inliers consistent with the instantiated model. These two steps are repeated for N minimal subsets of the data. The model having the largest number of inliers is chosen and refined by re-estimating it from the corresponding set of inliers.

Our framework deals with M meaningful matches, and usually M is small enough to test for all corresponding similarity or affine transformations. Hence, using the same ideas, an elementary algorithm would be as follows:

1. For each element in the set of M pairs of local frames corresponding to meaningful matches:
    a. Compute the associated transformation T;
    b. Apply T to all query local frames, and compute their distances to their corresponding scene local frames;
    c. Compute the number of inliers consistent with T, *i.e.* the pairs for which the distance is less than d pixels;
2. Choose the transformation T having the largest number of inliers;
3. Re-estimate T for all pairs of local frames determined as inliers (with a least squares method, for instance).

One can iterate this procedure on the set of outliers in order to find other (less dominant) transformations.

Even for this simple version of the algorithm, two problems arise: the choice of the distance threshold d, and the minimum number of inliers a model should have in order to be valid. The distance threshold d is usually chosen empirically. Otherwise, it can be chosen by considering a significance level α, corresponding to the probability that a point is an inlier [81], which requires hypothesizing a model for the distribution of distances. Concerning the minimum number of inliers to assess model validity, generally it is also fixed by means of arbitrary rules. It seems reasonable to us that this minimum number of inliers depends on the distance threshold, but as far as we know no effort has been done to establish this relation.

A.3 On the Negative Association of Multinomial Distributions

This section presents the notion of *negative association* (a strong notion of negative dependence) and summarizes some relevant consequences first reported by Joag-Dev and Proschan in [94]. Some proofs are also completed when they were just outlined in the original paper. The result is then applied to multinomial distributions.

Definition 20 (Negative association). A set $\mathcal{X} = \{X_1, \ldots, X_n\}$ of real random variables is said to be negatively associated (NA) if for every two disjoint index sets $I, J \subset \{1, \ldots, n\}$,

$$\mathbb{E}\left[f(X_i, i \in I)g(X_j, j \in J)\right] \leqslant \mathbb{E}\left[f(X_i, i \in I)\right] \cdot \mathbb{E}\left[g(X_j, j \in J)\right],$$

for all non-decreasing functions $f : \mathbb{R}^{\#I} \to \mathbb{R}$, $g : \mathbb{R}^{\#J} \to \mathbb{R}$ (a function $h : \mathbb{R}^k \to \mathbb{R}$ is said to be non-decreasing if $h(x_1, \ldots, x_k) \geqslant h(y_1, \ldots, y_k)$ whenever $x_1 \leqslant y_1, \ldots, x_k \leqslant y_k$).

Remark 7. Negative association is a natural generalization of negative correlation.

The negatively associated set $\mathcal{X} = \{X_1, \ldots, X_n\}$ verifies the following properties:

Property 1. For any non-decreasing functions f_i, $i \in \{1, \ldots, n\}$,

$$\mathbb{E}\left[\prod_{i=1}^{n} f_i(X_i)\right] \leqslant \prod_{i=1}^{n} \mathbb{E}\left[f_i(X_i)\right].$$

Proof. Define $f(x_1, \ldots, x_{n-1}) = \prod_{i=1}^{n-1} f_i(x_i)$ and $g(x_n) = f_n(x_n)$ for all $(x_1, \ldots, x_n) \in \mathbb{R}^n$. Since f and g are both non-decreasing, it follows from Definition 20 that

$$\mathbb{E}\left[\prod_{i=1}^{n} f_i(X_i)\right] \leqslant \mathbb{E}\left[\prod_{i=1}^{n-1} f_i(X_i)\right] \mathbb{E}\left[f_n(X_n)\right].$$

Using induction yields the desired result. $\square$

Property 2. For all $(x_1, \ldots, x_n) \in \mathbb{R}^n$,

$$\Pr\left(X_i \geqslant x_i \; \forall \, i \in \{1, \ldots, n\}\right) \leqslant \prod_{i=1}^{n} \Pr\left(X_i \geqslant x_i\right).$$

This follows immediately from Property 1 for $f_i(x) = \chi_{[x \geqslant x_i]}$, the indicator function of event $[x \geqslant x_i]$. The following property is obvious from Definition 20:

Property 3. Non-decreasing functions defined on disjoint subsets of a set of NA random variables are NA.

Property 4. The union of independent sets of NA random variables is NA.

Proof. Let $\mathbf{X}$ and $\mathbf{Y}$ be independent vectors such that for each one its components are sets of NA random variables. Let $(\mathbf{X}_1, \mathbf{X}_2)$ and $(\mathbf{Y}_1, \mathbf{Y}_2)$ denote arbitrary partitions of $\mathbf{X}$ and $\mathbf{Y}$ respectively. Hence, the vector $(\mathbf{X}, \mathbf{Y})$ is NA if and only if $\mathbb{E}\left[f(\mathbf{X}_1, \mathbf{Y}_1)g(\mathbf{X}_2, \mathbf{Y}_2)\right] \leqslant \mathbb{E}\left[f(\mathbf{X}_1, \mathbf{Y}_1)\right]\mathbb{E}\left[g(\mathbf{X}_2, \mathbf{Y}_2)\right]$. Now,

$$
\begin{aligned}
\mathbb{E}\left[f(\mathbf{X}_1, \mathbf{Y}_1)g(\mathbf{X}_2, \mathbf{Y}_2)\right] &= \mathbb{E}\left\{\mathbb{E}\left[f(\mathbf{X}_1, \mathbf{Y}_1)g(\mathbf{X}_2, \mathbf{Y}_2)|\mathbf{Y}_1, \mathbf{Y}_2\right]\right\} \\
&= \sum_{(y_1, y_2)} \mathbb{E}\left[f(\mathbf{X}_1, \mathbf{Y}_1)g(\mathbf{X}_2, \mathbf{Y}_2)|\mathbf{Y}_1 = y_1, \mathbf{Y}_2 = y_2\right] \\
&\qquad\qquad\qquad \cdot \Pr(\mathbf{Y}_1 = y1, \mathbf{Y}_2 = y2).
\end{aligned}
$$

Since $(\mathbf{X}_1, \mathbf{X}_2)$ and $(\mathbf{Y}_1, \mathbf{Y}_2)$ are independent, $\{f(\mathbf{X}_1, \mathbf{Y}_1)|\mathbf{Y}_1 = y_1, \mathbf{Y}_2 = y_2\}$ and $\{g(\mathbf{X}_2, \mathbf{Y}_2)|\mathbf{Y}_1 = y_1, \mathbf{Y}_2 = y_2\}$ are parametric functions of random vectors $\mathbf{X}_1$ and $\mathbf{X}_2$ respectively. Thus, because of the negative association of $\mathbf{X}$,

$$
\begin{aligned}
\mathbb{E}\left[f(\mathbf{X}_1, \mathbf{Y}_1)g(\mathbf{X}_2, \mathbf{Y}_2)|\mathbf{Y}_1 = y_1, \mathbf{Y}_2 = y_2\right] &\leqslant \\
\mathbb{E}\left[f(\mathbf{X}_1, \mathbf{Y}_1)|\mathbf{Y}_1 = y_1, \mathbf{Y}_2 = y_2\right] &\mathbb{E}\left[g(\mathbf{X}_2, \mathbf{Y}_2)|\mathbf{Y}_1 = y_1, \mathbf{Y}_2 = y_2\right].
\end{aligned}
$$

Hence,

$$
\mathbb{E}\left[f(\mathbf{X}_1, \mathbf{Y}_1)g(\mathbf{X}_2, \mathbf{Y}_2)\right] \leqslant \mathbb{E}\left\{\mathbb{E}\left[f(\mathbf{X}_1, \mathbf{Y}_1)|\mathbf{Y}_1, \mathbf{Y}_2\right]\mathbb{E}\left[g(\mathbf{X}_2, \mathbf{Y}_2)|\mathbf{Y}_1, \mathbf{Y}_2\right]\right\}
$$

Now, since the conditional expectations

$$
\mathbb{E}\left[f(\mathbf{X}_1, \mathbf{Y}_1)|\mathbf{Y}_1, \mathbf{Y}_2\right] \text{ and } \mathbb{E}\left[g(\mathbf{X}_2, \mathbf{Y}_2)|\mathbf{Y}_1, \mathbf{Y}_2\right]
$$

are respectively $\mathbf{Y}_1$ and $\mathbf{Y}_2$ measurable functions, it follows that

$$
\begin{aligned}
h_1(\mathbf{Y}_1) &\equiv \mathbb{E}\left[f(\mathbf{X}_1, \mathbf{Y}_1)|\mathbf{Y}_1, \mathbf{Y}_2\right] = \mathbb{E}\left[f(\mathbf{X}_1, \mathbf{Y}_1)|\mathbf{Y}_1\right], \\
h_2(\mathbf{Y}_2) &\equiv \mathbb{E}\left[g(\mathbf{X}_2, \mathbf{Y}_2)|\mathbf{Y}_1, \mathbf{Y}_2\right] = \mathbb{E}\left[g(\mathbf{X}_2, \mathbf{Y}_2)|\mathbf{Y}_2\right].
\end{aligned}
$$

Finally, using that $\mathbf{Y}$ is NA,

$$
\begin{aligned}
\mathbb{E}\left[f(\mathbf{X}_1, \mathbf{Y}_1)g(\mathbf{X}_2, \mathbf{Y}_2)\right] &\leqslant \mathbb{E}\left[h_1(\mathbf{Y}_1)h_2(\mathbf{Y}_2)\right] \\
&\leqslant \mathbb{E}\left[h_1(\mathbf{Y}_1)\right]\mathbb{E}\left[h_2(\mathbf{Y}_1)\right] \\
&= \mathbb{E}\left[f(\mathbf{X}_1, \mathbf{Y}_1)\right]\mathbb{E}\left[g(\mathbf{X}_2, \mathbf{Y}_2)\right]. \qquad \square
\end{aligned}
$$

These results yield the following proposition.

Proposition 14. *A random vector* $\mathbf{X} = (X_1, \ldots, X_n)$ *having a multinomial distribution of index M and parameter* $\mathbf{p} = (p_1, \ldots, p_n)$ *(denoted by* $\mathbf{X} \sim Mult(M, \mathbf{p})$*), is NA.*

Proof. The variable $\mathbf{X}$ can be decomposed as

$$\mathbf{X} = \sum_{k=1}^{M} \mathbf{Y}_k,$$

where each $\mathbf{Y}_k \sim Mult(1, \mathbf{p})$, and the $\mathbf{Y}_k$'s are mutually independent. Since, for all $k \in \{1, \ldots, M\}$, all elements in $\mathbf{Y}_k$ are zero except for one whose value is 1, vector $\mathbf{Y}_k$ is NA. Indeed, for all I, J disjoint subsets of $\{1, \ldots, n\}$, for all non-decreasing functions $f : \mathbb{R}^{\#I} \to \mathbb{R}$, $g : \mathbb{R}^{\#J} \to \mathbb{R}$,

$$\mathbb{E}\left[f(\mathbf{Y}_{k,i}, i \in I)g(\mathbf{Y}_{k,j}, j \in J)\right] \leqslant \mathbb{E}\left[f(\mathbf{Y}_{k,i}, i \in I)\right] \cdot \mathbb{E}\left[g(\mathbf{Y}_{k,j}, j \in J)\right]$$
$$\Leftrightarrow \mathbb{E}\left[(f(\mathbf{Y}_{k,i}, i \in I) - f(0, \ldots, 0))\,(g(\mathbf{Y}_{k,j}, j \in J) - g(0, \ldots, 0))\right]$$
$$\leqslant \mathbb{E}\left[f(\mathbf{Y}_{k,i}, i \in I) - f(0, \ldots, 0)\right] \cdot \mathbb{E}\left[g(\mathbf{Y}_{k,j}, j \in J) - g(0, \ldots, 0)\right].$$

The last inequality is true: the right member is non-negative because $f(\mathbf{Y}_{k,i}, i \in I) - f(0, \ldots, 0)$ and $g(\mathbf{Y}_{k,j}, j \in J) - g(0, \ldots, 0)$ are non-negative, and the left member is zero since $(f(\mathbf{Y}_{k,i}, i \in I) - f(0, \ldots, 0)$ and $g(\mathbf{Y}_{k,j}, j \in J) - g(0, \ldots, 0)$ cannot be non-zero at the same time.

Then, using Property 4, it follows that $\{\mathbf{Y}_1, \ldots, \mathbf{Y}_M\}$ is NA. Finally, for all $l \in \{1, \ldots, n\}$, $X_l = \sum_{k=1}^{M} \mathbf{Y}_{k,l}$ are non-decreasing functions defined on disjoint subsets of $\{\mathbf{Y}_1, \ldots, \mathbf{Y}_M\}$. This proves that $\mathbf{X}$ is NA (Property 3). $\square$

Remark 8. Applying Property A.3 to the random vector $\mathbf{X}$ proves Lem. 7, stated in Sect. 7.3.

Appendix B
Algorithms

B.1 LLD Method Summary

Input Image

STEP 1:
EXTRACTION

1.1. Level Lines Tree computation (Sect. 2.1)
1.2. Maximal Meaningful Boundaries selection.
   (Sect. 2.3.2). Active options:
   – Clean sub-curves in noise ($\mu = 1$) (Sect. 2.4.1.1)
   – Multiscale version (two levels, $N_s = 2$) (Sect. 2.5)
   – Local contrast (Sect. 2.6)

Maximal Boundaries

STEP 2:
ENCODING

2.1. Curves smoothing (Sect. 3.3)
2.2. Extraction of robust directions: bitangents
   and flat parts (Sect. 3.1)
2.3. Geometric normalization, similarity and affine
   invariant: global (Sect. 4.1.3) and local (Sect. 4.2)

Level Line Descriptors (LLDs)

Two sets of LLDs

STEP 3:
RECOGNITION

3.1. Background model construction: selection
   of statistically independent global and local
   features from LLDs (Sect. 5.2.1) and computation
   of empirical frequencies of distances between
   them (Sect. 5.2.1 and Eq. (5.9))
3.2. Given any two LLDs in the databases, compute
   their distance (Eq. (5.10)) and NFA (Def. 8)
3.3. Use Def. 9 to find ε-meaningful matching pairs

Meaningful Matching Pairs

STEP 4:
GROUPING

4.1. Build a background model on the set of similarities
   or affine transforms associated to the matching pairs
   (Sect. 8.3.2)
4.2. The transforms (similarity, Sect. 8.2.1; or affine,
   Sect. 8.2.2) associated with the matching pairs
   form a point data set
4.3. From this set a clustering tree is built (Sect. A.1.3)
   using the dissimilarity measures of Def. 17 (similarity
   case) or 18 (affine case)
4.4. Maximal groups are computed using the grouping
   algorithm in Sect. 7.4.1 and Def. 16

Maximal Meaningful Groups of Matches

B.2 Improved MSER Method Summary

Input Image

STEP 1:   1.1. Computation of the Tree of Shapes of the image
EXTRACTION   using FLST [135]
       1.2. Selection of Maximally Stable Extremal Regions
         (MSER) (Sect. 2.2)

MSERs

STEP 2:   2.1. Geometric global normalization (Sect. 4.1.3):
ENCODING   similarity and affine invariant

Improved MSERs descriptors

Two sets of improved MSER descriptors

STEP 3:   3.1. Background model construction: selection
RECOGNITION   of statistically independent global features
       from MSER descriptors (Sect. 5.2.1) and
       computation of empirical frequencies of distances
       between them (Sect. 5.2.1 and Eq. (5.9))
       3.2. Given any two MSER descriptors in the databases,
         compute their distance (Eq. (5.10)) and NFA (Def. 8)
       3.3. Use Def. 9 to find ε-meaningful matching pairs

Meaningful Matching Pairs

STEP 4:   4.1. Build a background model on the set of similarities
GROUPING   or affine transforms associated to the matching pairs
         (Sect. 8.3.2)
       4.2. The transforms (similarity, Sect. 8.2.1; or affine,
         Sect. 8.2.2) associated with the matching pairs
         form a point data set
       4.3. From this set a clustering tree is built (Sect. A.1.3)
         using the dissimilarity measures of Def. 17 (similarity
         case) or 18 (affine case)
       4.4. Maximal groups are computed using the grouping
         algorithm in Sect. 7.4.1 and Def. 16

Maximal Meaningful Groups of Matches

B.3 Improved SIFT Method Summary

Input Image

| **STEP 1:**
EXTRACTION | 1.1. Computation of keypoints: extrema of the
Laplacian in the image scale space. They are pairs
(position, scale). (Sect. 10.1.1, 10.1.2 and [114]) |

Keypoints

| **STEP 2:**
ENCODING | 2.1. Associate one or more orientations to each keypoint
(dominant gradient directions in the vicinity
of the point) (Sect. 10.1.3)
2.2. Associate an image patch to each keypoint
(region in the vicinity of the point)
2.3. Subdivide the square patch into 8×8 subsquares
2.4. Keep the orientation histograms inside each square
The orientations are weighted by gradient magnitude |

Improved SIFT descriptors: keypoint + scale + orientation + image patch
Two sets of improved SIFT descriptors

| **STEP 3:**
RECOGNITION | 3.1. Background model: the differences of
gradient directions between statistically
independent image points are uniformly
distributed in $(0, \pi)$ (Sect. 11.3.2)
3.2. Given any two keypoints in the databases
get their corresponding image patches (up to the
similarity transform between the keypoints)(Sect. 11.4)
3.3. Sort a set of statistically independent pixels
in the patches and compute the NFA of the matching
(Sect. 11.4)
3.4. Keep ε-meaningful matching pairs |

Meaningful Matching Pairs

| **STEP 4:**
GROUPING | 4.1. Build a background model on the set of similarities
associated to the matching pairs (Sect. 8.3.2)
4.2. The similarity transforms (Sect. 8.2.1) associated
with the matching pairs form a point data set
4.3. From this set a clustering tree is built (Sect. A.1.3)
using the dissimilarity measure of Def. 17 (similarity
case)
4.4. Maximal groups are computed using the grouping
algorithm in Sect. 7.4.1 and Def. 16 |

Maximal Meaningful Groups of Matches

References

1. Abbasi, S., Mokhtarian, F.: Retrieval of similar shapes under affine transformation. In: Proceedings of International Conference on Visual Information Systems, pp. 566–574. Amsterdam, The Netherlands (1999)
2. Almansa, A., Desolneux, A., Vamech, S.: Vanishing point detection without any a priori information. IEEE Transactions on Pattern Analysis and Machine Intelligence **25**(4), 502–507 (2003)
3. Alt, H., Guibas, L.: Discrete geometric shapes: Matching, interpolation, and approximation. In: J.R. Sack, J. Urrutia (eds.) Handbook of Computational Geometry, pp. 121–153. Elsevier Science Publishers (1999)
4. Alt, H., Knauer, C., Wenk, C.: Matching polygonal curves with respect to the Fréchet distance. In: Proceedings of the 18th International Symposium on Theoretical Aspects of Computer Science, pp. 63–74. Dresden, Germany (2001)
5. Alvarez, L., Guichard, F., Lions, P.L., Morel, J.M.: Axioms and fundamental equations of image processing: Multiscale analysis and P.D.E. Archive for Rational Mechanics and Analysis **16**(9), 200–257 (1993)
6. Alvarez, L., Mazorra, L., Santana, F.: Geometric invariant shape representations using morphological multiscale analysis and applications to shape representation. Journal of Mathematical Imaging and Vision **18**(2), 145–168 (2002)
7. Angenent, S., Sapiro, G., Tannenbaum, A.: On the affine heat flow for nonconvex curves. Journal of the American Mathematical Society (1998)
8. Ansari, N., Delp, E.J.: Partial shape recognition: A landmark-based approach. IEEE Transactions on Pattern Analysis and Machine Intelligence **12**(5), 470–483 (1990)
9. Arnaud, N., Cavalier, F., Davier, M., Hello, P.: Detection of gravitational wave bursts by interferometric detectors. Physical review D **59**(8), 082,002–1 – 082,002–9 (1999)
10. Asada, H., Brady, M.: The curvature primal sketch. IEEE Transactions on Pattern Analysis and Machine Intelligence **8**(1), 2–14 (1986)
11. Åström, K.: Affine and projective normalization of planar curves and regions. In: Proceedings of European Conference on Computer Vision, vol. 2, pp. 439–448. Stockholm, Sweden (1994)
12. Åström, K.: Fundamental limitations on projective invariants of planar curves. IEEE Transactions on Pattern Analysis and Machine Intelligence **17**(1), 77–81 (1995)
13. Attneave, F.: Some informational aspects of visual perception. Psychological review **61**(3), 183–193 (1954)
14. Ballard, D.: Generalizing the Hough transform to detect arbitrary shapes. Pattern Recognition **13**(2), 111–122 (1981)
15. Ballester, C., Caselles, V., Monasse, P.: The tree of shapes of an image. ESAIM: Control, Optimisation and Calculus of Variations **9**, 1–18 (2003)

16. Barles, G., Souganidis, P.: Convergence of approximation schemes for fully nonlinear second order equations. Asymptotic Analysis **4**, 271–283 (1991)
17. Barlow, H.: What is the computational goal of the neocortex? Large-scale neuronal theories of the brain. MIT Press, Cambridge, MA (1994)
18. Basri, R., Costa, L., Geiger, D., Jacobs, D.: Determining the similarity of deformable shapes. Vision Research **38**(15-16), 2365–2385 (1998)
19. Baumberg, A.: Reliable feature matching across widely separated views. In: Proceedings of the International Conference on Computer Vision and Pattern Recognition, pp. 774–781. Washington DC, USA (2004)
20. Beg, M.F., Miller, M., Trouvé, A., Younes, L.: Computing large deformation metric mappings via geodesic flows of diffeomorphisms. Internation Journal of Computer Vision **61**(2), 139–157 (2005)
21. Belongie, S., Carson, C., Greenspan, H., Malik, J.: Color- and texture-based image segmentation using the Expectation-Maximization algorithm and its application to content-based image retrieval. In: Proceedings of the International Conference on Computer Vision, pp. 675–682. Mumbai, India (1998)
22. Belongie, S., Malik, J., Puzicha, J.: Shape matching and object recognition using shape contexts. IEEE Transactions on Pattern Analysis and Machine Intelligence **24**(5), 509–522 (2002)
23. Bienenstock, E., Geman, S., Potter, D.: Compositionality, MDL priors, and object recognition. In: M. Mozera, M. Jordan, T. Petsche (eds.) Advances in Neural Information Processing Systems 9. MIT Press (1998)
24. Bock, H.: On some significance tests in cluster analysis. Journal of Classification **2**, 77–108 (1985)
25. Bottou, L., Bengio, Y.: Convergence properties of the k-means algorithms. In: G. Tesario, D. Touretzky (eds.) Advances in Neural Information Processing Systems, vol. 7, pp. 585–592. Denver, Colorado, USA (1995)
26. Calinski, T., Harabasz, J.: A dendrite method for cluster analysis. Communications in statistics **3**(1), 1–27 (1974)
27. Canny, J.: A variational approach to edge detection. In: National Conference on Artificial Intelligence, pp. 54–58. Washington DC, USA (1983)
28. Canny, J.: A computational approach to edge detection. IEEE Transactions on Pattern Analysis and Machine Intelligence **8**(6), 679–698 (1986)
29. Cao, F.: Geometric Curve Evolution and Image Processing, *Lecture Notes in Mathematics*, vol. 1805. Springer Verlag (2003)
30. Cao, F.: Good continuations in digital images. In: Proceeding of International Conference on Computer Vision, vol. 1, pp. 440–447. Nice, France (2003)
31. Cao, F.: Application of the Gestalt principles to the detection of good continuations and corners in image level lines. Computing and Visualisation in Science **7**(1), 3–13 (2004)
32. Cao, F., Bouthemy, P.: A general criterion for image similarity detection. Technical Report IRISA 1732 (2005)
33. Cao, F., Delon, J., Desolneux, A., Musé, P., Sur, F.: A unified framework for detecting groups and application to shape recognition. Journal of Mathematical Imaging and Vision **27**(2), 91–119 (2007)
34. Cao, F., Moisan, L.: A geometrical scheme for curve evolution driven by curvature. SIAM Journal of Numerical Analysis **39**(2), 624–646 (2000)
35. Cao, F., Musé, P., Sur, F.: Extracting meaningful curves from images. Journal of Mathematical Imaging and Vision **22**(2-3), 159–181 (2005)
36. Caselles, V., Coll, B., Morel, J.M.: A Kanizsa program. Progress in Nonlinear Differential Equations and their Applications **25**, 35–55 (1996)
37. Caselles, V., Coll, B., Morel, J.M.: Topographic maps and local contrast changes in natural images. International Journal of Computer Vision **33**(1), 5–27 (1999)
38. Chapple, P., Bertilone, D., Caprari, R., Newsam, G.: Stochastic model-based processing for detection of small targets in non-gaussian natural imagery. IEEE Transactions on Image Processing **10**(4), 554–564 (2001)

39. Cohen, F., Huang, Z., Yang, Z.: Invariant matching and identification of curves using B-Splines curve representation. IEEE Transactions on Image Processing **4**(1), 1–10 (1995)
40. Cohignac, T.: Reconnaissance de formes planes. Ph.D. thesis, Ceremade, Université Paris IX Dauphine (1994)
41. Cohignac, T., Lopez, C., Morel, J.M.: Integral and local affine invariant parameter and application to shape recognition. In: International Conference on Pattern Recognition, pp. A:164–168. Jerusalem, Israel (1994)
42. Cortadellas, J., Amat, J., de la Torre, F.: Robust normalization of silhouettes for recognition applications. Pattern Recognition Letters **25**(5), 591–601 (2004)
43. Cox, J., Karron, D.: Digital Morse theory (1998). Manuscript available at http://www.casi.net/D.DMT/D.Overview/AcademicPressPaper14-03
44. Day, W., Edelsbrunner, H.: Efficient algorithms for agglomerative hierarchical clustering methods. Journal of Classification **1**(1), 7–24 (1984)
45. Debled-Rennesson, I., Rémy, J.L., Rouyer-Degli, J.: Segmentation of discrete curves into fuzzy segments. Tech. Rep. 4989, INRIA (2003)
46. Dempster, A., Laird, N., Rubin, D.: Maximum likelihood from incomplete data via EM algorithm. Journal of the Royal Statistical Society, Series B **39**, 1–38 (1977)
47. Deriche, R.: Using Canny's criteria to derive a recursively implemented optimal edge detector. International Journal of Computer Vision **1**(2), 167–187 (1987)
48. Deriche, R., Faugeras, O.: Tracking line segments. Image and Vision Computing **8**(4), 261–270 (1990)
49. Desolneux, A., Ladjal, S., Moisan, L., Morel, J.M.: Dequantizing image orientation. IEEE Transactions on Image Processing **11**(10), 1129–1140 (2002)
50. Desolneux, A., Moisan, L., Morel, J.M.: Meaningful alignments. International Journal of Computer Vision **40**(1), 7–23 (2000)
51. Desolneux, A., Moisan, L., Morel, J.M.: Edge detection by Helmholtz principle. Journal of Mathematical Imaging and Vision **14**(3), 271–284 (2001)
52. Desolneux, A., Moisan, L., Morel, J.M.: Computational gestalts and perception thresholds. Journal of Physiology - Paris **97**(2-3), 311–322 (2003)
53. Desolneux, A., Moisan, L., Morel, J.M.: A grouping principle and four applications. IEEE Transactions on Pattern Analysis and Machine Intelligence **25**(4), 508–513 (2003)
54. Desolneux, A., Moisan, L., Morel, J.M.: Gestalt Theory and Image Analysis, a Probabilistic Approach. Interdisciplinary Applied Mathematics series, Springer Verlag (2007). Preprint available at http://www.cmla.ens-cachan.fr/Utilisateurs/morel/lecturenote.pdf
55. Devijver, P., Kittler, J.: Pattern recognition - A statistical approach. Prentice Hall (1982)
56. Donoser, M., Bischof, H.: Efficient Maximally Stable Extremal Region (MSER) Tracking. Proceedings of the 2006 IEEE Computer Society Conference on Computer Vision and Pattern Recognition-Volume 1 pp. 553–560 (2006)
57. Dryden, I.: General shape and registration analysis. Tech. rep., University of Leeds, Department of Statistics (1996)
58. Dubes, R..C.: How many clusters are best? – an experiment. Pattern Recognition **20**(6), 645–663 (1987)
59. Duda, R., Hart, P.: Pattern Classification and Scene Analysis. John Wiley and Sons (1973)
60. Duda, R., Hart, P., Stork, D.: Pattern Classification. John Wiley and Sons (2000). 2nd edition
61. Dudani, S., Breeding, K., McGhee, R.: Aircraft identification by moment invariants. IEEE Transactions on Computers **26**(1), 39–46 (1977)
62. Dudek, G., Tsotsos, J.: Shape representation and recognition from multiscale curvature. Computer Vision and Image Understanding **2**(68), 170–189 (1997)
63. Faugeras, O., Keriven, R.: Some recent results on the projective evolution of 2d curves. In: Proceedings of IEEE International Conference on Image Processing, vol. 3, pp. 13–16. Washington DC, USA (1995)
64. Fischler, M., Bolles, R.: Random sample consensus: A paradigm for model fitting with applications to image analysis and automated cartography. Communications of the Association for Computing Machinery **24**(6), 381–395 (1981)

65. Fischler, M., Bolles, R.: Perceptual organization and curve partitioning. IEEE Transactions on pattern analysis and machine intelligence **8**(1), 100–105 (1986)

66. Frosini, P., Landi, C.: Size theory as a topological tool for computer vision. Pattern Recognition and Image Analysis **9**(4), 596–603 (1999)

67. Frosini, P., Landi, C.: Size functions and formal series. Applicable Algebra in Engineering, Communication and Computing **12**, 327–349 (2001)

68. Gdalyahu, Y., Weinshall, D.: Flexible syntactic matching of curves and its application to automatic hierarchical classification of silhouettes. IEEE Transactions on Pattern Analysis and Machine Intelligence **21**(12), 1312–1328 (1999)

69. Geman, S., McClure, D.: Statistical methods for tomographic image reconstruction. Bulletin of the International Statistical Institute **52**, 5–21 (1987)

70. Giraudon, G.: Chaînage efficace de contour. Tech. Rep. 0605, INRIA (1987)

71. Golub, G., Loan, C.V.: Matrix Computations. Johns Hopkins University Press (1989)

72. Gordon, A.: Null models in cluster validation. In: W. Gaul, D. Pfeifer (eds.) From Data to Knowledge: Theoretical and Practical Aspects of Classification, Data Analysis, and Knowledge Organization, pp. 32–44. Springer Verlag (1996)

73. Gordon, A.: Classification. Monographs on Statistics and Applied Probability 82, Chapman & Hall (1999)

74. Gousseau, Y.: Comparaison de la composition de deux images, et application à la recherche automatique. In: proceedings of GRETSI 2003. Paris, France (2003)

75. Grimson, W., Huttenlocher, D.: On the sensitivity of the Hough transform for object recognition. IEEE Transactions on Pattern Analysis and Machine Intelligence **12**(3), 255–274 (1990)

76. Grimson, W., Huttenlocher, D.: On the verification of hypothesized matches in model-based recognition. IEEE Transactions on Pattern Analysis and Machine Intelligence **13**(12), 1201–1213 (1991)

77. Guigues, L.: Modèles multi-échelles pour la segmentation d'images. Ph.D. thesis, Université de Cergy-Pontoise (2003)

78. Guy, G., Medioni, G.: Inferring global perceptual contours from local features. International Journal on Computer Vision **20**(1), 113–133 (1996)

79. Haralick, R.: Digital step edges from zero crossing of second directional derivatives. IEEE Transactions on Pattern Analysis and Machine Intelligence **6**(1), 58–68 (1984)

80. Harris, C., Stephens, M.: A combined corner and edge detector. In: 4th Alvey Vision Conference, Manchester, pp. 189–192 (1988)

81. Hartley, R., Zisserman, A.: Multiple View Geometry in Computer Vision. Cambridge University Press (2000)

82. Hastie, T., Tibshirani, R., Friedman, J.: The Elements of Statistical Learning Data Mining, Inference, and Prediction. Springer Series in Statistics (2001)

83. Helmholtz, H.: Treatise on Physiological Optics. Dover, New York (1962 (first published in 1867))

84. Hoeffding, W.: Probability Inequalities for Sums of Bounded Random Variables. Journal of the American Statistical Association **58**(301), 13–30 (1963)

85. Hough, P.: Methods and means for recognizing complex patterns (1962). U.S. Patent 3,069,654

86. Hu, M.: Visual pattern recognition by moments invariants. IRE Transactions on Information Theory **8**, 179–187 (1962)

87. Huang, Z., Cohen, F.: Affine-invariant B-spline moments for curve matching. IEEE Transactions on Image Processing **5**(10), 1473–1480 (1996)

88. Huttenlocher, D., Klanderman, G., Rucklidge, W.: Comparing images using the Hausdorff distance. IEEE Transactions on Pattern Analysis and Machine Intelligence **15**(9), 850–863 (1993)

89. Huttenlocher, D., Ullman, S.: Object recognition using alignment. In: International Conference of Computer Vision, pp. 267–291. London, UK (1987)

90. Hyvärinen, A.: Survey on independent component analysis. Neural Computing Surveys **2**, 94–128 (1999)

91. Illingworth, J., Kittler, J.: A survey of the Hough transform. Computer Vision, Graphics, and Image Processing **44**(1), 87–116 (1988)

92. Jacobs, D.: Robust and efficient detection of salient convex groups. IEEE Transactions on Pattern Analysis and Machine Intelligence **18**(1), 23–37 (1996)

93. Jain, A.K., Murty, M.N., Flynn, P.J.: Data clustering: a review. ACM Computing Surveys **31**(3), 264–323 (1999)

94. Joag-Dev, K., Proschan, F.: Negative association of random variables, with applications. Annals of Statistics **11**(1), 286–295 (1983)

95. Kanizsa, G.: Organization in Vision: Essays on Gestalt Perception. Praeger (1979)

96. Kaufman, L., Rousseeuw, P.J.: Finding groups in data: an introduction to cluster analysis. John Wiley and Sons (1990)

97. Khalil, M., Bayoumi, M.: A dyadic wavelet affine invariant function for 2d shape recognition. IEEE Transactions on Pattern Analysis and Machine Intelligence **23**(10) (2001)

98. Koenderink, J.: The structure of images. Biological Cybernetics **50**, 363–370 (1984)

99. Koepfler, G., Moisan, L.: Geometric multiscale representation of numerical images. In: Second International Conference on Scale Space Theories in Computer Vision, *Lecture Notes in Computer Science*, vol. 1682, pp. 339–350. Springer, Corfu, Greece (1999)

100. Krzyzak, A., Leung, S., Suen, C.: Reconstruction of two-dimensional patterns from Fourier descriptors. Machine Vision and Applications **2**, 123–140 (1989)

101. Lamdan, Y., Schwartz, J., Wolfson, H.: Object recognition by affine invariant matching. In: Proceedings of IEEE International Conference on Computer Vision and Pattern Recognition, pp. 335–344. Ann Arbor, Michigan, U.S.A. (1988)

102. Lamdan, Y., Wolfson, H.: Geometric hashing: a general and efficient model-based recognition scheme. In: Proceedings of IEEE International Conference on Computer Vision, pp. 238–249. Tampa, Florida, USA (1988)

103. Lance, G., Williams, W.: A general theory of classificatory sorting strategies. I. hierarchical systems. Computer Journal **9**, 373–370 (1967)

104. Lin, C., Chellappa, R.: Classification of partial 2-d shapes using Fourier descriptors. In: Proceedings of the International Conference on Computer Vision and Pattern Recognition, pp. 344–350. Miami Beach, Florida, USA (1986)

105. Lindeberg, T.: Feature detection with automatic scale selection. International Journal of Computer Vision **30**(2), 77–116 (1998)

106. Lindenbaum, M.: An integrated model for evaluating the amount of data required for reliable recognition. IEEE Trans. Pattern Analysis Machine Intelligence **19**(11), 1251–1264 (1997)

107. Lindenbaum, M., Bruckstein, A.: On recursive, $O(N)$ partitioning of a digitized curve into digital straight segments. Transactions on Pattern Analysis and Machine Intelligence **15**(9) (1993)

108. Lisani, J.: Shape based automatic images comparison. Ph.D. thesis, Université Paris 9 Dauphine, France (2001)

109. Lisani, J., Moisan, L., Monasse, P., Morel, J.M.: On the theory of planar shape. SIAM Multiscale Modeling and Simulation **1**(1), 1–24 (2003)

110. Lisani, J., Monasse, P., Rudin, L.: Fast shape extraction and applications. Tech. Rep. 2001-16, CMLA, ENS Cachan (2001)

111. Loncaric, S.: A survey of shape analysis techniques. Pattern Recognition **31**(8), 983–1001 (1998)

112. Lowe, D.: Perceptual Organization and Visual Recognition. Kluwer Academic Publisher (1985)

113. Lowe, D.: Object recognition from local scale-invariant features. In: Proceedings of IEEE International Conference on Computer Vision, pp. 1150–1157. Corfu, Greece (1999)

114. Lowe, D.: Distinctive image features from scale-invariant key points. International Journal of Computer Vision **60**(2), 91–110 (2004)

115. MacQueen, J.B.: Some methods for classification and analysis of multivariate observations. In: Proceedings of the 5th Berkeley Symposium on Mathematical Statistics and Probability, vol. 1, pp. 63–74 (1967)

116. Marr, D.: Vision. Freeman Publishers (1982)
117. Marr, D., Hildreth, E.: Theory of edge detection. Proceeding of the Royal Society of London **B-207**, 187–207 (1980)
118. Matas, J., Chum, O., Urban, M., Pajdla, T.: Robust wide baseline stereo from maximally stable extremal regions. Image and Vision Computing **22**(10), 761–767 (2004)
119. Matheron, G.: Random Sets and Integral Geometry. John Wiley and Sons (1975)
120. Mikolajczyk, K., Schmid: A performance evaluation of local descriptors. IEEE Transactions on Pattern Analysis and Machine Intelligence **27**(10), 1615–1630 (2005)
121. Mikolajczyk, K., Schmid, C.: Scale & affine invariant interest point detectors. International Journal of Computer Vision **60**(1), 63–86 (2004)
122. Mikolajczyk, K., Tuytelaars, T., Schmid, C., Zisserman, A., Matas, J., Schaffalitzky, F., Kadir, T., Van Gool, L.: A comparison of affine region detectors. International Journal of Computer Vision **65**(1-2), 43–72 (2005)
123. Miller, M., Trouvé, A., Younes, L.: On the metrics and Euler-Lagrange equations of computational anatomy. Annual Review of Biomedical Engineering **4**, 375–405 (2002)
124. Miller, M., Trouvé, A., Younes, L.: Geodesic shooting for computational anatomy. Journal of Mathematical Imaging and Vision **24**(2), 209–228 (2006)
125. Milligan, G., Cooper, M.: An examination of procedures for determining the number of clusters in a data set. Psychometrika **50**(2), 159–179 (1985)
126. Milnor, J.: Topology from the Differentiable Viewpoint. Princeton University Press (1997)
127. Moisan, L.: Affine plane curve evolution: A fully consistent scheme. IEEE Transactions on Image Processing **7**(3), 411–420 (1998)
128. Moisan, L., Stival, B.: A probabilistic criterion to detect rigid point matches between two images and estimate the fundamental matrix. International Journal of Computer Vision **57**(3), 201–218 (2004)
129. Mokhtarian, F.: Silhouette-based isolated object recognition through curvature scale space. IEEE Transactions on Pattern Analysis and Machine Intelligence **17**(5), 539–544 (1995)
130. Mokhtarian, F., Abbasi, S., Kittler, J.: Efficient and robust retrieval by shape content through curvature scale space. In: Proceedings of International Workshop on Image Databases and MultiMedia Search, pp. 35–42. Amsterdam, The Netherlands (1996)
131. Mokhtarian, F., Abbasi, S., Kittler, J.: Robust and efficient shape indexing through curvature scale space. In: Proceedings of British Machine Vision Conference, pp. 53–62. Edinburgh, UK (1996)
132. Mokhtarian, F., Mackworth, A.: A theory of multiscale, curvature-based shape representation for planar curves. IEEE Transactions on Pattern Analysis and Machine Intelligence **14**(8), 789–805 (1992)
133. Monasse, P.: Contrast invariant image registration. In: Proceedings of International Conference on Acoustics, Speech and Signal Processing, vol. 6, pp. 3221–3224. Phoenix, Arizona, USA (1999)
134. Monasse, P.: Représentation morphologique d'images numériques et application au recalage, morphological representation of digital images and application to registration. Ph.D. thesis, Université Paris 9 Dauphine, France (2000)
135. Monasse, P., Guichard, F.: Fast computation of a contrast invariant image representation. IEEE Transactions on Image Processing **9**(5), 860–872 (2000)
136. Monasse, P., Guichard, F.: Scale-space from a level lines tree. Journal of Visual Communication and Image Representation **11**, 224–236 (2000)
137. Morel, J.M., Solimini, S.: Variational Methods in Image Segmentation. Birkhauser (1995)
138. Murphy-Chutorian, E., Trivedi, M.: N-tree Disjoint-Set Forests for Maximally Stable Extremal Regions. In: Proceedings of the British Machine Vision Conference. Edinburgh, United Kingdom (2006)
139. Musé, P., Sur, F., Cao, F., Gousseau, Y.: Unsupervised thresholds for shape matching. In: Proceedings of IEEE International Conference on Image Processing. Barcelona, Spain (2003)
140. Musé, P., Sur, F., Cao, F., Gousseau, Y., Morel, J.M.: An A Contrario Decision Method for Shape Element Recognition. International Journal of Computer Vision **69**(3), 295–315 (2006)

141. Musé, P., Sur, F., Morel, J.M.: Sur les seuils de reconnaissance des formes. Traitement du Signal **20**(3), 279–294 (2003)
142. Niethammer, M., Betelu, S., Sapiro, G., Tannenbaum, A., Giblin, P.: Area-based medial axis of planar curves. International Journal of Computer Vision **60**(3), 203–224 (2004)
143. Obdržálek, S., Matas, J.: Local affine frames for image retrieval. In: CIVR'02: Proceedings of International Conference The Challenge of Image and Video Retrieval, pp. 318–327. Springer-Verlag (2002)
144. Olson, C., Huttenlocher, D.: Automatic target recognition by matching oriented edge pixels. IEEE Transactions on Image Processing **6**(12), 103–113 (1997)
145. Orrite, C., Herrero, J.: Shape matching of partially occluded curves invariant under projective transformation. Computer Vision and Image Understing **93**(1), 34–64 (2004)
146. Papoulis, A., Bertran, M.: Digital filtering and prolate functions. IEEE Transactions on Circuits and Systems **19**(6), 674–681 (1972)
147. Pennec, X.: Toward a generic framework for recognition based on uncertain geometric features. Videre: Journal of Computer Vision Research **1**(2), 58–87 (1998)
148. Persoon, E., Fu, K.: Shape discrimination using Fourier descriptors. SMC **7**(3), 170–179 (1977)
149. Poor, H.: An Introduction to Signal Detection and Estimation, 2nd edn. Springer Texts in Electrical Engineering. Springer Verlag (1994)
150. Rabin, J., Gousseau, Y., Delon, J.: A contrario matching of local descriptors. Tech. Rep. hal-00168285, Ecole Nationale Supérieure des Télécommunications, Paris, France (2007)
151. Rothwell, C.: Object Recognition Through Invariant Indexing. Oxford Science Publications (1995)
152. Rothwell, C., Zisserman, A., Forsyth, D., Mundy, J.: Planar object recognition using projective shape representation. International Journal of Computer Vision **16**, 57–99 (1995)
153. Rubin, E.: Visuell wahrgenommene Figuren. Copenhagen, Gyldendals (1921)
154. Salembier, P., Serra, J.: Flat zones filtering, connected operators, and filters by reconstruction. IEEE Transactions on Image Processing **4**(8), 1153–1160 (1995)
155. Sapiro, G., Tannenbaum, A.: Affine invariant scale-space. International Journal of Computer Vision **11**(1), 25–44 (1993)
156. Sato, J., Cipolla, R.: Quasi-invariant parameterisations and matching of curves in images. International Journal of Computer Vision **28**(2), 117–138 (1998)
157. Schmid, C., G., Dorko, Lazebnik, S., Mikolajczyk, K., Ponce., J.: Pattern recognition with local invariant features. In: C. Chen, e. P.S.P Wang (eds.) Handbook of Pattern Recognition and Computer Vision. World Scientific Publishing Co. (2005). 3rd edition
158. Schmid, C., Mohr, R.: Combining greyvalue invariants with local constraints for object recognition. In: Proc. CVPR96, pp. 872–877. San Francisco, California (1996)
159. Schmid, C., Mohr, R.: Local greyvalue invariants for image retrieval. IEEE Transaction on Pattern Analysis and Machine Intelligence **19**(5), 530–535 (1997)
160. Sclaroff, S., Pentland, A.: Modal matching for correspondence and recognition. IEEE Transactions on Pattern Analysis and Machine Intelligence **17**(6), 545–561 (1995)
161. Serra, J.: Image Analysis and Mathematical Morphology. Academic Press (1982)
162. Shen, D., Ip, H.: Discriminative wavelet shape descriptors for recognition of 2-d patterns. Pattern Recognition **32**(8), 151–165 (1999)
163. Sivic, J., Zisserman, A.: Video Google: a text retrieval approach to object matching in videos. In: Ninth IEEE International Conference on Computer Vision, pp. 1470–1477 (2003)
164. Sklansky, J., Gonzalez, V.: Fast polygonal approximation of digitized curves. Pattern Recognition **12**, 327–331 (1980)
165. Small, C.: The Statistical Theory of Shapes. Springer Verlag (1996)
166. Song, Y.: A Top-down Algorithm for Computation of Level Line Trees. IEEE Transactions on Image Processing **16**(8), 2107–2116 (2007)
167. Song, Y., Zhang, A.: Analyzing scenery images by monotonic tree. Multimedia Systems **8**(6), 495–511 (2003)
168. Stewart, C.: MINPRAN: a new robust estimator for computer vision. IEEE Transactions on Pattern Analysis and Machine Intelligence **17**(10), 925–938 (1995)

169. Stockman, G., Kopstein, S., Benett, S.: Matching images to models for registration and object detection via clustering. IEEE Transactions on Pattern Analysis and Machine Intelligence **4**(3), 229–241 (1982)
170. Tan, P., Steinbach, M., Kumar, V.: Introduction to Data Mining. Addison-Wesley (2005)
171. de la Torre, F., Black, M.: A framework for robust subspace learning. International Journal of Computer Vision **54**(1-2-3), 117–142 (2003)
172. Veltkamp, R.: Shape matching: similarity measures and algorithms. In: Proceedings of International Conference on Shape Modeling and Applications, pp. 188–197. Genova, Italy (2001)
173. Veltkamp, R., Hagedoorn, M.: State-of-the-art in shape matching. In: M. Lew (ed.) Principles of Visual Information Retrieval, vol. 19. Springer Verlag (2001)
174. Veltkamp, R., Tanase, M.: Content-based image retrieval systems: A survey. Tech. Rep. UU-CS-2000-34, Utrecht University (2000)
175. Venters, C.C., Cooper, M.D.: A review of content-based image retrieval systems. Tech. Rep. jtap-054, University of Manchester, UK (2000)
176. Ward, J.H.J.: Hierarchical grouping to optimize an objective function. Journal of the American Statistical Association **58**(2), 236–244 (1963)
177. Watson, G., Watson, S.: Detection of unusual events in intermittent non-gaussian images using multiresolution background models. Optical Engineering **35**(11), 3159–3171 (1996)
178. Weiss, I.: Noise-resistant invariants of curves. IEEE Transactions on Pattern Analysis and Machine Intelligence **15**(9), 943–948 (1993)
179. Wertheimer, M.: Untersuchungen zur Lehre der Gestalt, II. Psychologische Forschung **4**, 301–350 (1923). Translation published as Laws of Organization in Perceptual Forms, in Ellis, W. (1938). A source book of Gestalt psychology (pp. 71-88). Routledge & Kegan Paul
180. Winter, A., Nastar, C.: Differential feature distribution maps for image segmentation and region queries in image databases. In: CBAIVL Workshop at Conference on Computer Vision and Pattern Recognition. Fort Collins, Colorado, USA (1999)
181. Witkin, A.: Scale space filtering. In: Proceedings of International Joint Conference on Artificial Intelligence, pp. 1019–1021. Karlsruhe, Germany (1983)
182. Wolfson, H.: Model-based object recognition by Geometric Hashing. In: Proceedings of the European Conference on Computer Vision, pp. 526–536. Lecture Notes in Computer Vision 427, Springer Verlag, Antibes, France (1990)
183. Wolfson, H.: On curve matching. IEEE Transactions on Pattern Analysis and Machine Intelligence **12**(5), 483–489 (1990)
184. Wolfson, H., Rigoutsos, I.: Geometric hashing: an overview. IEEE Computational Science & Engineering **4**(4), 10–21 (1997)
185. Zahn, C., Roskies, R.: Fourier descriptors for plane closed curves. IEEE Transactions on Computers **C-21**(3), 269–281 (1972)

Index

Symbols

ε-meaningful boundary, 19
ε-meaningful shape element matching, necessary and sufficient condition, 84
ε-meaningful shape matching, 83
a contrario detection, 129
a contrario detection, 2, 18, 19, 22, 26, 43, 57, 83, 89, 92, 130, 132, 149, 232

A

affine basis, 66, 70
affine curve shortening, 53
affine distortion, 41, 53, 74
affine invariance, 73, 75, 76
affine invariant encoding, 70, 73
affine invariant local frames, 71
affine invariant moments, 77
affine invariant normalization, 61
affine invariant shape elements, 11
affine invariant shape recognition, 237
affine invariant shape retrieval, 236
affine invariant smoothing, 10
affine morphological scale space, 76
affine scale space, 9, 53, 54, 59, 75
affine semi-local encoding, 71
affine shape encoding, 59
affine shape normalization by Cholesky method, 62
affine transformation, 76, 153, 154
asymptotic estimate (of the minimal number of points in a cluster), 135
Attneave, 4, 6, 7, 9, 10, 18, 76, 77

B

background model, vi, 2, 3, 11, 81, 83, 86, 89, 129, 131, 136, 149, 151, 153, 157, 159, 164, 233
background model for shape distances, 85
background point process, 131
binomial law, 132
bitangent line, 67, 75, 77
blur, 6, 19, 27, 82

C

clustering, 158
contrast, 28, 33
contrast (local), 31, 71
contrast (of boundaries), 18
contrast along level lines, 27
contrast change, 6
contrast changes (invariance to), 4, 6, 10, 34
contrast distribution, 29
contrast histogram, 31
contrast invariance, 15
contrast invariant information, 49
curvature, 53, 59
curvature motion, 53

D

dendrogram, 141, 142, 228
dissimilarity, 157
dissimilarity measure of two transformations, 156

E

edge detection, 32, 57, 75, 92

edge detector, 58
expectation of the number of ε-meaningful
    curves, 23
expectation of the number of ε-meaningful
    regions, 133
expectation of the number of ε-meaningful
    matches, 83
expected number of ε-meaningful pairs of
    regions, 138

F

figure-background problem, 5, 6, 39, 74
figure-background problem, 5
flat parts of a circle, 44

G

geometric hashing, 236
gestalt, 9, 32, 164
global affine invariant normalization, 61
global encoding, 66, 73
global normalization (geometric), 65
grouping, 2, 142, 151, 155, 159, 164

H

Helmholtz principle, vi, 1, 2, 26, 43, 90
hierarchical clustering, 12, 141, 228–231
Hough transform, 56, 57, 161, 235, 236
hyperrectangle, 132, 140

I

independence, 4, 23, 42, 87, 89, 90, 151, 157
indicator function, 62
indivisibility, necessary condition, 140
invariance of a normalized shape by Cholesky
    method, 63

J

Jordan curve, 16
Jordan level lines in images, 16

K

Kanizsa, 4, 5, 10

L

level line, 6–8, 10, 11, 16, 18–21, 24, 28, 30,
    32, 34, 35, 39, 43, 47, 49, 57, 61, 62, 65,
    67, 70, 73, 88, 90, 92, 153, 255, 256

LLD, 6–8, 10–12, 39, 61, 69, 71, 73, 81, 93,
    151, 153, 158, 167, 185
local encoding, 11, 33, 69, 73, 87, 89

M

maximal meaningful boundary, 20
maximal ε-meaningful group, 142
maximal meaningful alignments, 58
maximal meaningful boundary, 19, 20, 29
maximal meaningful cluster, 135, 143, 159
maximal meaningful group, 256
maximal monotone section, 20
meaningful match, 85
merging condition of two clusters, 139
merging (of clusters), 129, 136, 141, 142, 228,
    230
monotone section in the level line tree,
    definition, 20
MSER, 6, 10, 11, 116, 151, 185, 207, 208
multiscale (boundaries), 26
multiscale representation, 10
mutinomial law, 136

N

negative association of random variables, 239
NFA, 1, 3, 19, 20, 22, 27, 30, 31, 83, 90,
    134–136, 139, 141, 142, 159
NFA of a match of shape elements, 3
NFA of a cluster region, 133
NFA of a match of shape elements, 84
NFA of a pair of cluster regions, 136
NFA of pair of matching shape elements, 83
noise, 6, 7, 10, 15, 18, 24–28, 33, 43, 47, 49,
    56, 62, 63, 73–77, 90–92, 235
normalization of curves, consistency, 66

O

occlusion, 4, 6, 10, 12, 39, 74, 75

P

perspective, 6, 8, 73, 74
projective transformations, 73

R

RANSAC, 237
robust direction of a shape, 68

S

shape element, 151

shape element, 2, 3, 9–12, 33, 35, 61, 69, 81–85, 92, 129–132, 151, 153
  see also LLD, 7
shape element, general definition, 39
SIFT, vi, 6–8, 10–12, 106, 151, 185, 209
similarity invariance, 73
smoothing, 7–9, 27, 33, 39, 49, 54, 55, 57, 61, 73, 76
stroboscopic effect, 159, 179, 202, 223
sub-sampling, 10

T

texture, 32
topographic map, 15
tree of level lines, 20, 21, 29, 30, 33–35, 72
tree structure of point data set, 141, 159
trinomial and binomial (inequality), 140

W

Wertheimer, 4, 10, 164

Lecture Notes in Mathematics

For information about earlier volumes
please contact your bookseller or Springer
LNM Online archive: springerlink.com

Vol. 1766: H. Hennion, L. Hervé, Limit Theorems for Markov Chains and Stochastic Properties of Dynamical Systems by Quasi-Compactness (2001)

Vol. 1767: J. Xiao, Holomorphic Q Classes (2001)

Vol. 1768: M. J. Pflaum, Analytic and Geometric Study of Stratified Spaces (2001)

Vol. 1769: M. Alberich-Carramiñana, Geometry of the Plane Cremona Maps (2002)

Vol. 1770: H. Gluesing-Luerssen, Linear Delay-Differential Systems with Commensurate Delays: An Algebraic Approach (2002)

Vol. 1771: M. Émery, M. Yor (Eds.), Séminaire de Probabilités 1967-1980. A Selection in Martingale Theory (2002)

Vol. 1772: F. Burstall, D. Ferus, K. Leschke, F. Pedit, U. Pinkall, Conformal Geometry of Surfaces in S^4 (2002)

Vol. 1773: Z. Arad, M. Muzychuk, Standard Integral Table Algebras Generated by a Non-real Element of Small Degree (2002)

Vol. 1774: V. Runde, Lectures on Amenability (2002)

Vol. 1775: W. H. Meeks, A. Ros, H. Rosenberg, The Global Theory of Minimal Surfaces in Flat Spaces. Martina Franca 1999. Editor: G. P. Pirola (2002)

Vol. 1776: K. Behrend, C. Gomez, V. Tarasov, G. Tian, Quantum Comohology. Cetraro 1997. Editors: P. de Bartolomeis, B. Dubrovin, C. Reina (2002)

Vol. 1777: E. García-Río, D. N. Kupeli, R. Vázquez-Lorenzo, Osserman Manifolds in Semi-Riemannian Geometry (2002)

Vol. 1778: H. Kiechle, Theory of K-Loops (2002)

Vol. 1779: I. Chueshov, Monotone Random Systems (2002)

Vol. 1780: J. H. Bruinier, Borcherds Products on O(2,1) and Chern Classes of Heegner Divisors (2002)

Vol. 1781: E. Bolthausen, E. Perkins, A. van der Vaart, Lectures on Probability Theory and Statistics. Ecole d' Eté de Probabilités de Saint-Flour XXIX-1999. Editor: P. Bernard (2002)

Vol. 1782: C.-H. Chu, A. T.-M. Lau, Harmonic Functions on Groups and Fourier Algebras (2002)

Vol. 1783: L. Grüne, Asymptotic Behavior of Dynamical and Control Systems under Perturbation and Discretization (2002)

Vol. 1784: L. H. Eliasson, S. B. Kuksin, S. Marmi, J.-C. Yoccoz, Dynamical Systems and Small Divisors. Cetraro, Italy 1998. Editors: S. Marmi, J.-C. Yoccoz (2002)

Vol. 1785: J. Arias de Reyna, Pointwise Convergence of Fourier Series (2002)

Vol. 1786: S. D. Cutkosky, Monomialization of Morphisms from 3-Folds to Surfaces (2002)

Vol. 1787: S. Caenepeel, G. Militaru, S. Zhu, Frobenius and Separable Functors for Generalized Module Categories and Nonlinear Equations (2002)

Vol. 1788: A. Vasil'ev, Moduli of Families of Curves for Conformal and Quasiconformal Mappings (2002)

Vol. 1789: Y. Sommerhäuser, Yetter-Drinfel'd Hopf algebras over groups of prime order (2002)

Vol. 1790: X. Zhan, Matrix Inequalities (2002)

Vol. 1791: M. Knebusch, D. Zhang, Manis Valuations and Prüfer Extensions I: A new Chapter in Commutative Algebra (2002)

Vol. 1792: D. D. Ang, R. Gorenflo, V. K. Le, D. D. Trong, Moment Theory and Some Inverse Problems in Potential Theory and Heat Conduction (2002)

Vol. 1793: J. Cortés Monforte, Geometric, Control and Numerical Aspects of Nonholonomic Systems (2002)

Vol. 1794: N. Pytheas Fogg, Substitution in Dynamics, Arithmetics and Combinatorics. Editors: V. Berthé, S. Ferenczi, C. Mauduit, A. Siegel (2002)

Vol. 1795: H. Li, Filtered-Graded Transfer in Using Noncommutative Gröbner Bases (2002)

Vol. 1796: J.M. Melenk, hp-Finite Element Methods for Singular Perturbations (2002)

Vol. 1797: B. Schmidt, Characters and Cyclotomic Fields in Finite Geometry (2002)

Vol. 1798: W.M. Oliva, Geometric Mechanics (2002)

Vol. 1799: H. Pajot, Analytic Capacity, Rectifiability, Menger Curvature and the Cauchy Integral (2002)

Vol. 1800: O. Gabber, L. Ramero, Almost Ring Theory (2003)

Vol. 1801: J. Azéma, M. Émery, M. Ledoux, M. Yor (Eds.), Séminaire de Probabilités XXXVI (2003)

Vol. 1802: V. Capasso, E. Merzbach, B. G. Ivanoff, M. Dozzi, R. Dalang, T. Mountford, Topics in Spatial Stochastic Processes. Martina Franca, Italy 2001. Editor: E. Merzbach (2003)

Vol. 1803: G. Dolzmann, Variational Methods for Crystalline Microstructure – Analysis and Computation (2003)

Vol. 1804: I. Cherednik, Ya. Markov, R. Howe, G. Lusztig, Iwahori-Hecke Algebras and their Representation Theory. Martina Franca, Italy 1999. Editors: V. Baldoni, D. Barbasch (2003)

Vol. 1805: F. Cao, Geometric Curve Evolution and Image Processing (2003)

Vol. 1806: H. Broer, I. Hoveijn. G. Lunther, G. Vegter, Bifurcations in Hamiltonian Systems. Computing Singularities by Gröbner Bases (2003)

Vol. 1807: V. D. Milman, G. Schechtman (Eds.), Geometric Aspects of Functional Analysis. Israel Seminar 2000-2002 (2003)

Vol. 1808: W. Schindler, Measures with Symmetry Properties (2003)

Vol. 1809: O. Steinbach, Stability Estimates for Hybrid Coupled Domain Decomposition Methods (2003)

Vol. 1810: J. Wengenroth, Derived Functors in Functional Analysis (2003)

Vol. 1811: J. Stevens, Deformations of Singularities (2003)

Vol. 1812: L. Ambrosio, K. Deckelnick, G. Dziuk, M. Mimura, V. A. Solonnikov, H. M. Soner, Mathematical Aspects of Evolving Interfaces. Madeira, Funchal, Portugal 2000. Editors: P. Colli, J. F. Rodrigues (2003)

Vol. 1813: L. Ambrosio, L. A. Caffarelli, Y. Brenier, G. Buttazzo, C. Villani, Optimal Transportation and its Applications. Martina Franca, Italy 2001. Editors: L. A. Caffarelli, S. Salsa (2003)

Vol. 1814: P. Bank, F. Baudoin, H. Föllmer, L.C.G. Rogers, M. Soner, N. Touzi, Paris-Princeton Lectures on Mathematical Finance 2002 (2003)

Vol. 1815: A. M. Vershik (Ed.), Asymptotic Combinatorics with Applications to Mathematical Physics. St. Petersburg, Russia 2001 (2003)

Vol. 1816: S. Albeverio, W. Schachermayer, M. Talagrand, Lectures on Probability Theory and Statistics. Ecole d'Eté de Probabilités de Saint-Flour XXX-2000. Editor: P. Bernard (2003)

Vol. 1817: E. Koelink, W. Van Assche (Eds.), Orthogonal Polynomials and Special Functions. Leuven 2002 (2003)

Vol. 1818: M. Bildhauer, Convex Variational Problems with Linear, nearly Linear and/or Anisotropic Growth Conditions (2003)

Vol. 1819: D. Masser, Yu. V. Nesterenko, H. P. Schlickewei, W. M. Schmidt, M. Waldschmidt, Diophantine Approximation. Cetraro, Italy 2000. Editors: F. Amoroso, U. Zannier (2003)

Vol. 1820: F. Hiai, H. Kosaki, Means of Hilbert Space Operators (2003)

Vol. 1821: S. Teufel, Adiabatic Perturbation Theory in Quantum Dynamics (2003)

Vol. 1822: S.-N. Chow, R. Conti, R. Johnson, J. Mallet-Paret, R. Nussbaum, Dynamical Systems. Cetraro, Italy 2000. Editors: J. W. Macki, P. Zecca (2003)

Vol. 1823: A. M. Anile, W. Allegretto, C. Ringhofer, Mathematical Problems in Semiconductor Physics. Cetraro, Italy 1998. Editor: A. M. Anile (2003)

Vol. 1824: J. A. Navarro González, J. B. Sancho de Salas, $\mathscr{C}^{\infty}$ – Differentiable Spaces (2003)

Vol. 1825: J. H. Bramble, A. Cohen, W. Dahmen, Multiscale Problems and Methods in Numerical Simulations, Martina Franca, Italy 2001. Editor: C. Canuto (2003)

Vol. 1826: K. Dohmen, Improved Bonferroni Inequalities via Abstract Tubes. Inequalities and Identities of Inclusion-Exclusion Type. VIII, 113 p, 2003.

Vol. 1827: K. M. Pilgrim, Combinations of Complex Dynamical Systems. IX, 118 p, 2003.

Vol. 1828: D. J. Green, Gröbner Bases and the Computation of Group Cohomology. XII, 138 p, 2003.

Vol. 1829: E. Altman, B. Gaujal, A. Hordijk, Discrete-Event Control of Stochastic Networks: Multimodularity and Regularity. XIV, 313 p, 2003.

Vol. 1830: M. I. Gil', Operator Functions and Localization of Spectra. XIV, 256 p, 2003.

Vol. 1831: A. Connes, J. Cuntz, E. Guentner, N. Higson, J. E. Kaminker, Noncommutative Geometry, Martina Franca, Italy 2002. Editors: S. Doplicher, L. Longo (2004)

Vol. 1832: J. Azéma, M. Émery, M. Ledoux, M. Yor (Eds.), Séminaire de Probabilités XXXVII (2003)

Vol. 1833: D.-Q. Jiang, M. Qian, M.-P. Qian, Mathematical Theory of Nonequilibrium Steady States. On the Frontier of Probability and Dynamical Systems. IX, 280 p, 2004.

Vol. 1834: Yo. Yomdin, G. Comte, Tame Geometry with Application in Smooth Analysis. VIII, 186 p, 2004.

Vol. 1835: O.T. Izhboldin, B. Kahn, N.A. Karpenko, A. Vishik, Geometric Methods in the Algebraic Theory of Quadratic Forms. Summer School, Lens, 2000. Editor: J.-P. Tignol (2004)

Vol. 1836: C. Năstăsescu, F. Van Oystaeyen, Methods of Graded Rings. XIII, 304 p, 2004.

Vol. 1837: S. Tavaré, O. Zeitouni, Lectures on Probability Theory and Statistics. Ecole d'Eté de Probabilités de Saint-Flour XXXI-2001. Editor: J. Picard (2004)

Vol. 1838: A.J. Ganesh, N.W. O'Connell, D.J. Wischik, Big Queues. XII, 254 p, 2004.

Vol. 1839: R. Gohm, Noncommutative Stationary Processes. VIII, 170 p, 2004.

Vol. 1840: B. Tsirelson, W. Werner, Lectures on Probability Theory and Statistics. Ecole d'Eté de Probabilités de Saint-Flour XXXII-2002. Editor: J. Picard (2004)

Vol. 1841: W. Reichel, Uniqueness Theorems for Variational Problems by the Method of Transformation Groups (2004)

Vol. 1842: T. Johnsen, A. L. Knutsen, K_3 Projective Models in Scrolls (2004)

Vol. 1843: B. Jefferies, Spectral Properties of Noncommuting Operators (2004)

Vol. 1844: K.F. Siburg, The Principle of Least Action in Geometry and Dynamics (2004)

Vol. 1845: Min Ho Lee, Mixed Automorphic Forms, Torus Bundles, and Jacobi Forms (2004)

Vol. 1846: H. Ammari, H. Kang, Reconstruction of Small Inhomogeneities from Boundary Measurements (2004)

Vol. 1847: T.R. Bielecki, T. Björk, M. Jeanblanc, M. Rutkowski, J.A. Scheinkman, W. Xiong, Paris-Princeton Lectures on Mathematical Finance 2003 (2004)

Vol. 1848: M. Abate, J. E. Fornaess, X. Huang, J. P. Rosay, A. Tumanov, Real Methods in Complex and CR Geometry, Martina Franca, Italy 2002. Editors: D. Zaitsev, G. Zampieri (2004)

Vol. 1849: Martin L. Brown, Heegner Modules and Elliptic Curves (2004)

Vol. 1850: V. D. Milman, G. Schechtman (Eds.), Geometric Aspects of Functional Analysis. Israel Seminar 2002-2003 (2004)

Vol. 1851: O. Catoni, Statistical Learning Theory and Stochastic Optimization (2004)

Vol. 1852: A.S. Kechris, B.D. Miller, Topics in Orbit Equivalence (2004)

Vol. 1853: Ch. Favre, M. Jonsson, The Valuative Tree (2004)

Vol. 1854: O. Saeki, Topology of Singular Fibers of Differential Maps (2004)

Vol. 1855: G. Da Prato, P.C. Kunstmann, I. Lasiecka, A. Lunardi, R. Schnaubelt, L. Weis, Functional Analytic Methods for Evolution Equations. Editors: M. Iannelli, R. Nagel, S. Piazzera (2004)

Vol. 1856: K. Back, T.R. Bielecki, C. Hipp, S. Peng, W. Schachermayer, Stochastic Methods in Finance, Bressanone/Brixen, Italy, 2003. Editors: M. Fritelli, W. Runggaldier (2004)

Vol. 1857: M. Émery, M. Ledoux, M. Yor (Eds.), Séminaire de Probabilités XXXVIII (2005)

Vol. 1858: A.S. Cherny, H.-J. Engelbert, Singular Stochastic Differential Equations (2005)

Vol. 1859: E. Letellier, Fourier Transforms of Invariant Functions on Finite Reductive Lie Algebras (2005)

Vol. 1860: A. Borisyuk, G.B. Ermentrout, A. Friedman, D. Terman, Tutorials in Mathematical Biosciences I. Mathematical Neurosciences (2005)

Vol. 1861: G. Benettin, J. Henrard, S. Kuksin, Hamiltonian Dynamics – Theory and Applications, Cetraro, Italy, 1999. Editor: A. Giorgilli (2005)

Vol. 1862: B. Helffer, F. Nier, Hypoelliptic Estimates and Spectral Theory for Fokker-Planck Operators and Witten Laplacians (2005)

Vol. 1863: H. Führ, Abstract Harmonic Analysis of Continuous Wavelet Transforms (2005)

Vol. 1864: K. Efstathiou, Metamorphoses of Hamiltonian Systems with Symmetries (2005)

Vol. 1865: D. Applebaum, B.V. R. Bhat, J. Kustermans, J. M. Lindsay, Quantum Independent Increment Processes I. From Classical Probability to Quantum Stochastic Calculus. Editors: M. Schürmann, U. Franz (2005)

Vol. 1866: O.E. Barndorff-Nielsen, U. Franz, R. Gohm, B. Kümmerer, S. Thorbjørnsen, Quantum Independent Increment Processes II. Structure of Quantum Lévy Processes, Classical Probability, and Physics. Editors: M. Schürmann, U. Franz, (2005)

Vol. 1867: J. Sneyd (Ed.), Tutorials in Mathematical Biosciences II. Mathematical Modeling of Calcium Dynamics and Signal Transduction. (2005)

Vol. 1868: J. Jorgenson, S. Lang, $Pos_n(R)$ and Eisenstein Series. (2005)

Vol. 1869: A. Dembo, T. Funaki, Lectures on Probability Theory and Statistics. Ecole d'Eté de Probabilités de Saint-Flour XXXIII-2003. Editor: J. Picard (2005)

Vol. 1870: V.I. Gurariy, W. Lusky, Geometry of Müntz Spaces and Related Questions. (2005)

Vol. 1871: P. Constantin, G. Gallavotti, A.V. Kazhikhov, Y. Meyer, S. Ukai, Mathematical Foundation of Turbulent Viscous Flows, Martina Franca, Italy, 2003. Editors: M. Cannone, T. Miyakawa (2006)

Vol. 1872: A. Friedman (Ed.), Tutorials in Mathematical Biosciences III. Cell Cycle, Proliferation, and Cancer (2006)

Vol. 1873: R. Mansuy, M. Yor, Random Times and Enlargements of Filtrations in a Brownian Setting (2006)

Vol. 1874: M. Yor, M. Émery (Eds.), In Memoriam Paul-André Meyer - Séminaire de Probabilités XXXIX (2006)

Vol. 1875: J. Pitman, Combinatorial Stochastic Processes. Ecole d'Eté de Probabilités de Saint-Flour XXXII-2002. Editor: J. Picard (2006)

Vol. 1876: H. Herrlich, Axiom of Choice (2006)

Vol. 1877: J. Steuding, Value Distributions of L-Functions (2007)

Vol. 1878: R. Cerf, The Wulff Crystal in Ising and Percolation Models, Ecole d'Eté de Probabilités de Saint-Flour XXXIV-2004. Editor: Jean Picard (2006)

Vol. 1879: G. Slade, The Lace Expansion and its Applications, Ecole d'Eté de Probabilités de Saint-Flour XXXIV-2004. Editor: Jean Picard (2006)

Vol. 1880: S. Attal, A. Joye, C.-A. Pillet, Open Quantum Systems I, The Hamiltonian Approach (2006)

Vol. 1881: S. Attal, A. Joye, C.-A. Pillet, Open Quantum Systems II, The Markovian Approach (2006)

Vol. 1882: S. Attal, A. Joye, C.-A. Pillet, Open Quantum Systems III, Recent Developments (2006)

Vol. 1883: W. Van Assche, F. Marcellàn (Eds.), Orthogonal Polynomials and Special Functions, Computation and Application (2006)

Vol. 1884: N. Hayashi, E.I. Kaikina, P.I. Naumkin, I.A. Shishmarev, Asymptotics for Dissipative Nonlinear Equations (2006)

Vol. 1885: A. Telcs, The Art of Random Walks (2006)

Vol. 1886: S. Takamura, Splitting Deformations of Degenerations of Complex Curves (2006)

Vol. 1887: K. Habermann, L. Habermann, Introduction to Symplectic Dirac Operators (2006)

Vol. 1888: J. van der Hoeven, Transseries and Real Differential Algebra (2006)

Vol. 1889: G. Osipenko, Dynamical Systems, Graphs, and Algorithms (2006)

Vol. 1890: M. Bunge, J. Funk, Singular Coverings of Toposes (2006)

Vol. 1891: J.B. Friedlander, D.R. Heath-Brown, H. Iwaniec, J. Kaczorowski, Analytic Number Theory, Cetraro, Italy, 2002. Editors: A. Perelli, C. Viola (2006)

Vol. 1892: A. Baddeley, I. Bárány, R. Schneider, W. Weil, Stochastic Geometry, Martina Franca, Italy, 2004. Editor: W. Weil (2007)

Vol. 1893: H. Hanßmann, Local and Semi-Local Bifurcations in Hamiltonian Dynamical Systems, Results and Examples (2007)

Vol. 1894: C.W. Groetsch, Stable Approximate Evaluation of Unbounded Operators (2007)

Vol. 1895: L. Molnár, Selected Preserver Problems on Algebraic Structures of Linear Operators and on Function Spaces (2007)

Vol. 1896: P. Massart, Concentration Inequalities and Model Selection, Ecole d'Été de Probabilités de Saint-Flour XXXIII-2003. Editor: J. Picard (2007)

Vol. 1897: R. Doney, Fluctuation Theory for Lévy Processes, Ecole d'Été de Probabilités de Saint-Flour XXXV-2005. Editor: J. Picard (2007)

Vol. 1898: H.R. Beyer, Beyond Partial Differential Equations, On linear and Quasi-Linear Abstract Hyperbolic Evolution Equations (2007)

Vol. 1899: Séminaire de Probabilités XL. Editors: C. Donati-Martin, M. Émery, A. Rouault, C. Stricker (2007)

Vol. 1900: E. Bolthausen, A. Bovier (Eds.), Spin Glasses (2007)

Vol. 1901: O. Wittenberg, Intersections de deux quadriques et pinceaux de courbes de genre 1, Intersections of Two Quadrics and Pencils of Curves of Genus 1 (2007)

Vol. 1902: A. Isaev, Lectures on the Automorphism Groups of Kobayashi-Hyperbolic Manifolds (2007)

Vol. 1903: G. Kresin, V. Maz'ya, Sharp Real-Part Theorems (2007)

Vol. 1904: P. Giesl, Construction of Global Lyapunov Functions Using Radial Basis Functions (2007)

Vol. 1905: C. Prévôt, M. Röckner, A Concise Course on Stochastic Partial Differential Equations (2007)

Vol. 1906: T. Schuster, The Method of Approximate Inverse: Theory and Applications (2007)

Vol. 1907: M. Rasmussen, Attractivity and Bifurcation for Nonautonomous Dynamical Systems (2007)

Vol. 1908: T.J. Lyons, M. Caruana, T. Lévy, Differential Equations Driven by Rough Paths, Ecole d'Été de Probabilités de Saint-Flour XXXIV-2004 (2007)

Vol. 1909: H. Akiyoshi, M. Sakuma, M. Wada, Y. Yamashita, Punctured Torus Groups and 2-Bridge Knot Groups (I) (2007)

Vol. 1910: V.D. Milman, G. Schechtman (Eds.), Geometric Aspects of Functional Analysis. Israel Seminar 2004-2005 (2007)

Vol. 1911: A. Bressan, D. Serre, M. Williams, K. Zumbrun, Hyperbolic Systems of Balance Laws. Cetraro, Italy 2003. Editor: P. Marcati (2007)

Vol. 1912: V. Berinde, Iterative Approximation of Fixed Points (2007)
Vol. 1913: J.E. Marsden, G. Misiołek, J.-P. Ortega, M. Perlmutter, T.S. Ratiu, Hamiltonian Reduction by Stages (2007)
Vol. 1914: G. Kutyniok, Affine Density in Wavelet Analysis (2007)
Vol. 1915: T. Bıyıkoğlu, J. Leydold, P.F. Stadler, Laplacian Eigenvectors of Graphs. Perron-Frobenius and Faber-Krahn Type Theorems (2007)
Vol. 1916: C. Villani, F. Rezakhanlou, Entropy Methods for the Boltzmann Equation. Editors: F. Golse, S. Olla (2008)
Vol. 1917: I. Veselić, Existence and Regularity Properties of the Integrated Density of States of Random Schrödinger (2008)
Vol. 1918: B. Roberts, R. Schmidt, Local Newforms for GSp(4) (2007)
Vol. 1919: R.A. Carmona, I. Ekeland, A. Kohatsu-Higa, J.-M. Lasry, P.-L. Lions, H. Pham, E. Taflin, Paris-Princeton Lectures on Mathematical Finance 2004. Editors: R.A. Carmona, E. Çinlar, I. Ekeland, E. Jouini, J.A. Scheinkman, N. Touzi (2007)
Vol. 1920: S.N. Evans, Probability and Real Trees. Ecole d'Été de Probabilités de Saint-Flour XXXV-2005 (2008)
Vol. 1921: J.P. Tian, Evolution Algebras and their Applications (2008)
Vol. 1922: A. Friedman (Ed.), Tutorials in Mathematical BioSciences IV. Evolution and Ecology (2008)
Vol. 1923: J.P.N. Bishwal, Parameter Estimation in Stochastic Differential Equations (2008)
Vol. 1924: M. Wilson, Littlewood-Paley Theory and Exponential-Square Integrability (2008)
Vol. 1925: M. du Sautoy, L. Woodward, Zeta Functions of Groups and Rings (2008)
Vol. 1926: L. Barreira, V. Claudia, Stability of Nonautonomous Differential Equations (2008)
Vol. 1927: L. Ambrosio, L. Caffarelli, M.G. Crandall, L.C. Evans, N. Fusco, Calculus of Variations and Non-Linear Partial Differential Equations. Cetraro, Italy 2005. Editors: B. Dacorogna, P. Marcellini (2008)
Vol. 1928: J. Jonsson, Simplicial Complexes of Graphs (2008)
Vol. 1929: Y. Mishura, Stochastic Calculus for Fractional Brownian Motion and Related Processes (2008)
Vol. 1930: J.M. Urbano, The Method of Intrinsic Scaling. A Systematic Approach to Regularity for Degenerate and Singular PDEs (2008)
Vol. 1931: M. Cowling, E. Frenkel, M. Kashiwara, A. Valette, D.A. Vogan, Jr., N.R. Wallach, Representation Theory and Complex Analysis. Venice, Italy 2004. Editors: E.C. Tarabusi, A. D'Agnolo, M. Picardello (2008)
Vol. 1932: A.A. Agrachev, A.S. Morse, E.D. Sontag, H.J. Sussmann, V.I. Utkin, Nonlinear and Optimal Control Theory. Cetraro, Italy 2004. Editors: P. Nistri, G. Stefani (2008)
Vol. 1933: M. Petković, Point Estimation of Root Finding Methods (2008)
Vol. 1934: C. Donati-Martin, M. Émery, A. Rouault, C. Stricker (Eds.), Séminaire de Probabilités XLI (2008)
Vol. 1935: A. Unterberger, Alternative Pseudodifferential Analysis (2008)
Vol. 1936: P. Magal, S. Ruan (Eds.), Structured Population Models in Biology and Epidemiology (2008)
Vol. 1937: G. Capriz, P. Giovine, P.M. Mariano (Eds.), Mathematical Models of Granular Matter (2008)

Vol. 1938: D. Auroux, F. Catanese, M. Manetti, P. Seidel, B. Siebert, I. Smith, G. Tian, Symplectic 4-Manifolds and Algebraic Surfaces. Cetraro, Italy 2003. Editors: F. Catanese, G. Tian (2008)
Vol. 1939: D. Boffi, F. Brezzi, L. Demkowicz, R.G. Durán, R.S. Falk, M. Fortin, Mixed Finite Elements, Compatibility Conditions, and Applications. Cetraro, Italy 2006. Editors: D. Boffi, L. Gastaldi (2008)
Vol. 1940: J. Banasiak, V. Capasso, M.A.J. Chaplain, M. Lachowicz, J. Miękisz, Multiscale Problems in the Life Sciences. From Microscopic to Macroscopic. Będlewo, Poland 2006. Editors: V. Capasso, M. Lachowicz (2008)
Vol. 1941: S.M.J. Haran, Arithmetical Investigations. Representation Theory, Orthogonal Polynomials, and Quantum Interpolations (2008)
Vol. 1942: S. Albeverio, F. Flandoli, Y.G. Sinai, SPDE in Hydrodynamic. Recent Progress and Prospects. Cetraro, Italy 2005. Editors: G. Da Prato, M. Röckner (2008)
Vol. 1943: L.L. Bonilla (Ed.), Inverse Problems and Imaging. Martina Franca, Italy 2002 (2008)
Vol. 1944: A. Di Bartolo, G. Falcone, P. Plaumann, K. Strambach, Algebraic Groups and Lie Groups with Few Factors (2008)
Vol. 1945: F. Brauer, P. van den Driessche, J. Wu (Eds.), Mathematical Epidemiology (2008)
Vol. 1946: G. Allaire, A. Arnold, P. Degond, T.Y. Hou, Quantum Transport. Modelling, Analysis and Asymptotics. Cetraro, Italy 2006. Editors: N.B. Abdallah, G. Frosali (2008)
Vol. 1947: D. Abramovich, M. Mariño, M. Thaddeus, R. Vakil, Enumerative Invariants in Algebraic Geometry and String Theory. Cetraro, Italy 2005. Editors: K. Behrend, M. Manetti (2008)
Vol. 1948: F. Cao, J.-L. Lisani, J.-M. Morel, P. Musé, F. Sur, A Theory of Shape Identification (2008)
Vol. 1949: H.G. Feichtinger, B. Helffer, M.P. Lamoureux, N. Lerner, J. Toft, Pseudo-differential Operators. Cetraro, Italy 2006. Editors: L. Rodino, M.W. Wong (2008)
Vol. 1950: M. Bramson, Stability of Queueing Networks, Ecole d' Eté de Probabilités de Saint-Flour XXXVI-2006 (2008)

Recent Reprints and New Editions

Vol. 1702: J. Ma, J. Yong, Forward-Backward Stochastic Differential Equations and their Applications. 1999 – Corr. 3rd printing (2007)
Vol. 830: J.A. Green, Polynomial Representations of GL_n, with an Appendix on Schensted Correspondence and Littelmann Paths by K. Erdmann, J.A. Green and M. Schoker 1980 – 2nd corr. and augmented edition (2007)
Vol. 1693: S. Simons, From Hahn-Banach to Monotonicity (Minimax and Monotonicity 1998) – 2nd exp. edition (2008)
Vol. 470: R.E. Bowen, Equilibrium States and the Ergodic Theory of Anosov Diffeomorphisms. With a preface by D. Ruelle. Edited by J.-R. Chazottes. 1975 – 2nd rev. edition (2008)
Vol. 523: S.A. Albeverio, R.J. Høegh-Krohn, S. Mazzucchi, Mathematical Theory of Feynman Path Integral. 1976 – 2nd corr. and enlarged edition (2008)
Vol. 1764: A. Cannas da Silva, Lectures on Symplectic Geometry 2001 – Corr. 2nd printing (2008)

LECTURE NOTES IN MATHEMATICS   ⚞ **Springer**

Edited by J.-M. Morel, F. Takens, B. Teissier, P.K. Maini

Editorial Policy (for the publication of monographs)

1. Lecture Notes aim to report new developments in all areas of mathematics and their applications - quickly, informally and at a high level. Mathematical texts analysing new developments in modelling and numerical simulation are welcome.

   Monograph manuscripts should be reasonably self-contained and rounded off. Thus they may, and often will, present not only results of the author but also related work by other people. They may be based on specialised lecture courses. Furthermore, the manuscripts should provide sufficient motivation, examples and applications. This clearly distinguishes Lecture Notes from journal articles or technical reports which normally are very concise. Articles intended for a journal but too long to be accepted by most journals, usually do not have this "lecture notes" character. For similar reasons it is unusual for doctoral theses to be accepted for the Lecture Notes series, though habilitation theses may be appropriate.

2. Manuscripts should be submitted either to Springer's mathematics editorial in Heidelberg, or to one of the series editors. In general, manuscripts will be sent out to 2 external referees for evaluation. If a decision cannot yet be reached on the basis of the first 2 reports, further referees may be contacted: The author will be informed of this. A final decision to publish can be made only on the basis of the complete manuscript, however a refereeing process leading to a preliminary decision can be based on a pre-final or incomplete manuscript. The strict minimum amount of material that will be considered should include a detailed outline describing the planned contents of each chapter, a bibliography and several sample chapters.

   Authors should be aware that incomplete or insufficiently close to final manuscripts almost always result in longer refereeing times and nevertheless unclear referees' recommendations, making further refereeing of a final draft necessary.

   Authors should also be aware that parallel submission of their manuscript to another publisher while under consideration for LNM will in general lead to immediate rejection.

3. Manuscripts should in general be submitted in English. Final manuscripts should contain at least 100 pages of mathematical text and should always include

   – a table of contents;
   – an informative introduction, with adequate motivation and perhaps some historical remarks: it should be accessible to a reader not intimately familiar with the topic treated;
   – a subject index: as a rule this is genuinely helpful for the reader.

   For evaluation purposes, manuscripts may be submitted in print or electronic form, in the latter case preferably as pdf- or zipped ps-files. Lecture Notes volumes are, as a rule, printed digitally from the authors' files. To ensure best results, authors are asked to use the LaTeX2e style files available from Springer's web-server at:

   ftp://ftp.springer.de/pub/tex/latex/svmonot1/ (for monographs).

Additional technical instructions, if necessary, are available on request from: lnm@springer.com.

4. Careful preparation of the manuscripts will help keep production time short besides ensuring satisfactory appearance of the finished book in print and online. After acceptance of the manuscript authors will be asked to prepare the final LaTeX source files (and also the corresponding dvi-, pdf- or zipped ps-file) together with the final printout made from these files. The LaTeX source files are essential for producing the full-text online version of the book (see www.springerlink.com/content/110312 for the existing online volumes of LNM).

The actual production of a Lecture Notes volume takes approximately 12 weeks.

5. Authors receive a total of 50 free copies of their volume, but no royalties. They are entitled to a discount of 33.3% on the price of Springer books purchased for their personal use, if ordering directly from Springer.

6. Commitment to publish is made by letter of intent rather than by signing a formal contract. Springer-Verlag secures the copyright for each volume. Authors are free to reuse material contained in their LNM volumes in later publications: a brief written (or e-mail) request for formal permission is sufficient.

Addresses:
Professor J.-M. Morel, CMLA,
École Normale Supérieure de Cachan,
61 Avenue du Président Wilson, 94235 Cachan Cedex, France
E-mail: Jean-Michel.Morel@cmla.ens-cachan.fr

Professor F. Takens, Mathematisch Instituut,
Rijksuniversiteit Groningen, Postbus 800,
9700 AV Groningen, The Netherlands
E-mail: F.Takens@math.rug.nl

Professor B. Teissier, Institut Mathématique de Jussieu,
UMR 7586 du CNRS, Équipe "Géométrie et Dynamique",
175 rue du Chevaleret
75013 Paris, France
E-mail: teissier@math.jussieu.fr

For the "Mathematical Biosciences Subseries" of LNM:

Professor P.K. Maini, Center for Mathematical Biology,
Mathematical Institute, 24-29 St Giles,
Oxford OX1 3LP, UK
E-mail: maini@maths.ox.ac.uk

Springer, Mathematics Editorial I, Tiergartenstr. 17
69121 Heidelberg, Germany,
Tel.: +49 (6221) 487-8410
Fax: +49 (6221) 4876-8259
E-mail: lnm@springer.com

Druck: Krips bv, Meppel, Niederlande
Verarbeitung: Stürtz, Würzburg, Deutschland